Oxford Physics Series

Oxford Physics Series

---

1. F. N. H. Robinson: *Electromagnetism*
2. G. Lancaster: *D.c. and a.c. circuits*
3. D. J. E. Ingram: *Radiation and quantum physics*
4. J. A. R. Griffith: *Interactions of particles*
5. B. R. Jennings and V. J. Morris: *Atoms in contact*
6. G. K. T. Conn: *Atoms and their structure*

F. N. H. ROBINSON
CLARENDON LABORATORY, OXFORD

Electromagnetism

Clarendon Press · Oxford · 1973

Oxford University Press, Ely House, London W.1

GLASGOW NEW YORK TORONTO MELBOURNE WELLINGTON
CAPE TOWN IBADAN NAIROBI DAR ES SALAAM LUSAKA ADDIS ABABA
DELHI BOMBAY CALCUTTA MADRAS KARACHI LAHORE DACCA
KUALA LUMPUR SINGAPORE HONG KONG TOKYO

PAPERBACK ISBN 0 19 851801 3
CASEBOUND ISBN 0 19 851806 4

PRINTED IN GREAT BRITAIN BY
J. W. ARROWSMITH LTD., BRISTOL, ENGLAND

Editor's foreword

ELECTROMAGNETISM is universally recognized to be one of the most difficult subjects to teach in any undergraduate physics course—and also one of the most difficult topics on which to write a satisfactory textbook. It is nevertheless one of the essential core subjects of physics, and thorough understanding of its concepts is essential to many of the more advanced topics. For this reason a book on electromagnetism was considered to be an essential member of the 'core texts' of the Oxford Physics Series. It has been included at the beginning of the second-year level, and is designed to follow on from Lancaster's *d.c. and a.c. circuits*, which gives all the necessary background related to the mathematical solution of circuit problems.

Although the presentation of the material is, of necessity, rather concise, Dr. Robinson underlines the various applications of the theory presented in each chapter by sets of very well chosen and relevant problems. When considering the properties of magnetic and electric fields inside solid bodies, he specifically avoids ideas and concepts which are not related sufficiently to the real world of practice and experiment. This approach, which is to be found especially in Chapters 7 and 8, together with a careful analysis of the meanings and implications of the different terms there introduced, should prove very helpful, and add the interest of a fresh approach.

This text has been written to cover its own particular topic in a concise and coherent manner, and also to fit carefully into the integrated structure of the Oxford Physics Series. Introductory core texts cover *Radiation and quantum physics*, *Atoms and their structure*, *Interactions of particles*, *Atoms in contact*, *d.c. and a.c. circuits*, *Electromagnetism*, and the mathematics required to back up these books. These core texts relate closely to each other and lead on to second- and third-year topics in quantum mechanics, statistical mechanics, solid-state and surface physics, electronics, nuclear physics, and space physics. These later topics can be rearranged to suit particular course structures, and optional subjects can be linked in as appropriate. In this way the whole Series is designed to reflect and match the more flexible nature of the new physics courses that are being designed at the moment—to give variety of approach but, at the same time, an integrated and coherent total picture.

Preface

IN this short book about one of the central topics of physics we shall be dealing solely with the fundamental laws and ideas of electromagnetism. Applications to subjects such as electronics, optics, or atomic physics will be mentioned only in passing and even topics such as relativity, circuit theory, electron optics, thermal radiation, or experimental techniques are omitted. The emphasis is on basic principles and concepts rather than on mathematical techniques for solving problems. This does not, however, mean that the book is non-mathematical, far from it: electromagnetism is the most highly developed branch of theoretical physics and its most important laws and results can only be concisely expressed in mathematical language. In addition to an understanding of the use of complex numbers in a.c. circuit theory and the use of $\exp(j\omega t)$ to represent a sinusoidal oscillation of frequency $\omega/2\pi$ the reader will be expected to be familiar with vectors, the notions of line, surface, and volume integral and the concepts implicit in expressions such as $\nabla\phi$, $\nabla \wedge \boldsymbol{E}$, and $\nabla . \boldsymbol{B}$.

The text uses SI units but, because other units are still important in the laboratory and occur in the classic literature, some of the problems involve other units. We do not adopt the recommendation that the magnetic vector $\boldsymbol{H}$ be called the field, but rather follow the commoner and more rational usage in which the vector $\boldsymbol{B}$ is called the field. The term magnetic intensity is used for $\boldsymbol{H}$.

The book as a whole is only an introductory text, but it should prepare the reader for a more advanced course based on some of the texts quoted in the bibliography at the end. These texts, being preoccupied with the further development of the theory, tend to give only very cursory attention to the fundamental laws and concepts. The treatment of much of the material follows along conventional lines but Chapters 7 and 8, dealing with fields in matter, are less conventional. The usual elementary treatment of these topics is inherently inconsistent and ultimately leads to confusion—confusion which is apparent not only at the undergraduate level, but also in much of the research literature. The approach used here is relatively advanced and, as a result, these two chapters are rather long. This seemed necessary, for this important topic cannot be avoided and there seemed to be little point in giving a grossly oversimplified account containing the seeds of error.

There are problems at the end of each chapter and these should not be omitted. They are an integral part of the book. Finally the author would like to thank Dr G. A. Brooker for his incisive and helpful comments on the text.

F. N. H. ROBINSON

Contents

1. Electrostatics

Charge

ELECTRIFIED bodies exert forces on each other which may be repulsive or attractive. The forces act along the line between the bodies and fall off as the inverse square of their distance apart. Action and reaction balance. If an electrified body C is brought to a distance r from an electrified body A and the force exerted on it is F_{AC} while, when it is brought to the same distance r from a third electrified body B, the force is F_{BC}, then the ratio F_{AC}/F_{BC} is independent of the state of electrification of the test body C. These laws allow us to assign to each body a numerical measure q, its charge, such that the forces exerted and experienced by the body are proportional to q. By allowing q to take positive and negative values we can combine the whole series of observations in a single equation for the force $\boldsymbol{F}(0)$ exerted on a body of charge $q(0)$ at $\boldsymbol{r}(0)$ due to a body of charge $q(1)$ at $\boldsymbol{r}(1)$. If we let $\boldsymbol{r}(0, 1) = \boldsymbol{r}(0) - \boldsymbol{r}(1)$ be the vector of length $r(0, 1)$ from $\boldsymbol{r}(1)$ to $\boldsymbol{r}(0)$, this equation is

$$\boldsymbol{F}(0) = \frac{q(0)q(1)\boldsymbol{r}(0, 1)}{4\pi\epsilon_0 r^3(0, 1)}. \tag{1.1}$$

Like charges repel, unlike charges attract. The constant ϵ_0 is connected with a choice of units for the charge q. The charge q is, like mass in gravitational problems, first introduced as a constant coupling a body to a force field.

Electric field

The effects of several charges $q(1)$, $q(2)$ etc. at $\boldsymbol{r}(1)$, $\boldsymbol{r}(2)$ etc. on a charge $q(0)$ at $\boldsymbol{r}(0)$ are additive and the force on $q(0)$ is

$$\boldsymbol{F}(0) = q(0) \sum_{n=1}^{\text{etc.}} \frac{q(n)\boldsymbol{r}(0, n)}{4\pi\epsilon_0 r^3(0, n)}. \tag{1.2}$$

We define the electric field $\boldsymbol{E}(\boldsymbol{r})$ at $\boldsymbol{r}$ so that the force $\boldsymbol{F}$ on a stationary body of charge q at $\boldsymbol{r}$ is

$$\boldsymbol{F} = q\boldsymbol{E}(\boldsymbol{r}). \tag{1.3}$$

The field is a vector and it is convenient to discuss its properties in terms of lines of $\boldsymbol{E}$ or lines of electric force. In a moving liquid the streamlines or trajectories of particles of the liquid, are lines everywhere parallel to the local velocity vector $\boldsymbol{v}$. The velocity is not constant on a streamline. However, if the fluid flows through a constriction the streamlines crowd together as the velocity increases at the constriction. The density of streamlines is proportional to the magnitude of the local velocity. Lines of electric force behave analogously. They are parallel to $\boldsymbol{E}$ at every point and the density of lines is proportional to the electric field strength. The electric field of a positive point charge is shown in Fig. 1.1 and we see how the density of lines diminishes with distance. The convention is that lines of force emanate from positive charge.

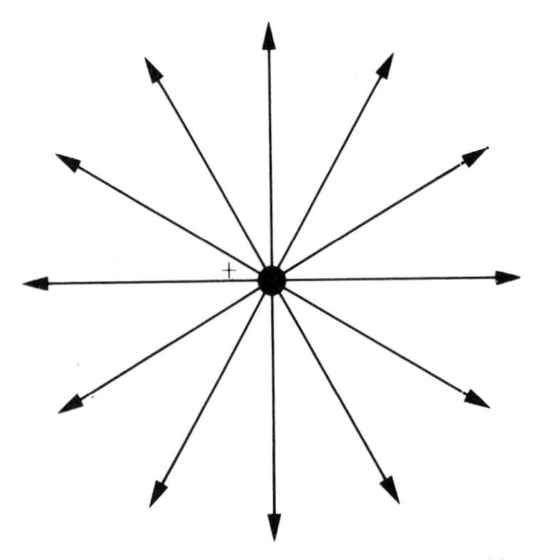

FIG. 1.1. Lines of electric field emanating from a positive charge.

Conservation of charge

Charge can be transported by the bodily motion of charged bodies or it can flow in conductors without any apparent mass motion. We find, with great precision, that if a region A gains charge δq from a region B then B loses an exactly equal charge, thus the algebraic sum of all the charges in an isolated region is constant or conserved. If a current of value I leaves a region A the charge Q in A decreases at a rate $\dot{Q} = -I$. We define the vector current density $\boldsymbol{J}$ so that the current crossing a plane surface element of area $\mathrm{d}\boldsymbol{S}$ in the direction of its positive normal is $\boldsymbol{J}.\mathrm{d}\boldsymbol{S}$. If ρ is the charge density, the charge within a volume V is the volume integral $\int_V \rho\, \mathrm{d}V$ and, if S is the surface bounding V, with positive outward normal, the current leaving V is the closed surface integral $\oint_S \boldsymbol{J}.\mathrm{d}\boldsymbol{S}$. Thus charge conservation implies the equation

$$\int_{\text{volume}} \dot{\rho}\, \mathrm{d}V + \oint_{\text{surface}} \boldsymbol{J}.\mathrm{d}\boldsymbol{S} = 0. \tag{1.4a}$$

The divergence of a vector $\boldsymbol{\nabla}.\boldsymbol{J}$ is defined so that for an infinitesimal volume δV the flux of $\boldsymbol{J}$ out of δV i.e. $\oint \boldsymbol{J}.\mathrm{d}\boldsymbol{S}$ is $\delta V \boldsymbol{\nabla}.\boldsymbol{J}$. If we take the volume in eqn

(1.4a) to be infinitesimal this gives the differential statement of the law of charge conservation

$$\nabla . \boldsymbol{J} + \dot{\rho} = 0. \tag{1.4b}$$

The notion that a force coupling constant such as q is conserved is an important component of the basic laws of electromagnetism.

The electrostatic potential

The gradient of a scalar $\psi(\boldsymbol{r})$ is defined so that

$$\mathrm{d}\psi = \psi(\boldsymbol{r}+\mathrm{d}\boldsymbol{r}) - \psi(\boldsymbol{r}) = \mathrm{d}\boldsymbol{r} . \nabla\psi(\boldsymbol{r}).$$

If

$$\psi(\boldsymbol{r}(0)) = \frac{1}{r(0, n)},$$

the gradient obtained by differentiating with respect to the terminal point $\boldsymbol{r}(0)$ of the vector $\boldsymbol{r}(0, n)$ is

$$\nabla_0\left(\frac{1}{r(0, n)}\right) = -\frac{\boldsymbol{r}(0, n)}{r^3(0, n)}. \tag{1.5}$$

(Notice that

$$\nabla_n\left(\frac{1}{r(0, n)}\right)$$

obtained by differentiation with respect to $\boldsymbol{r}(n)$ is of the opposite sign.) This can be used in conjunction with (1.2) and (1.3), which give the electric field as

$$\boldsymbol{E}(\boldsymbol{r}(0)) = \sum_n \frac{q(n)\boldsymbol{r}(0, n)}{4\pi\epsilon_0 r^3(0, n)}. \tag{1.6}$$

to express the field as

$$\boldsymbol{E}(\boldsymbol{r}(0)) = -\nabla_0\phi(\boldsymbol{r}(0)), \tag{1.7}$$

where

$$\phi(\boldsymbol{r}(0)) = \sum_n \frac{q(n)}{4\pi\epsilon_0 r(0, n)}. \tag{1.8}$$

The work that must be done by externally applied forces to take a charge q from $\boldsymbol{r}(1)$ to $\boldsymbol{r}(2)$ by a specified path is the line integral

$$W = -\int_{r(1)}^{r(2)} q\boldsymbol{E}(\boldsymbol{r}) . \mathrm{d}\boldsymbol{r}, \tag{1.9}$$

and, using (1.7) and the definition of $\nabla\phi$, we find that

$$W = \int_{\phi(r(1))}^{\phi(r(2))} q\, \mathrm{d}\phi = q\{\phi(\boldsymbol{r}(2)) - \phi(\boldsymbol{r}(1))\}. \tag{1.10}$$

This not only shows that $\phi(\boldsymbol{r})$ is the potential energy of a unit charge at $\boldsymbol{r}$ but also that W is independent of the path. It follows that the work done round a closed path is zero. The electrostatic field is a conservative field and for any closed path the line integral

$$\oint \boldsymbol{E}.\mathrm{d}\boldsymbol{r} = 0. \tag{1.11}$$

The curl of a vector $\boldsymbol{\nabla} \wedge \boldsymbol{E}$ is defined so that the line integral of $\boldsymbol{E}$ round a closed path, which is the circumference of a plane element of area $\mathrm{d}\boldsymbol{S}$, is $(\boldsymbol{\nabla} \wedge \boldsymbol{E}).\mathrm{d}\boldsymbol{S}$. If eqn (1.11) is applied to an infinitesimal path it yields the differential equation

$$\boldsymbol{\nabla} \wedge \boldsymbol{E} = 0. \tag{1.12}$$

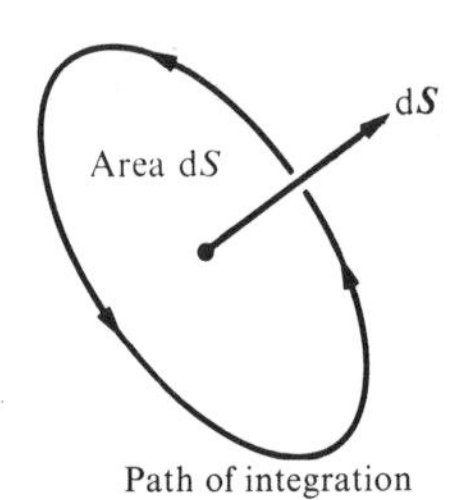

FIG. 1.2. Line and area elements used to define $\nabla \wedge \boldsymbol{E}$.

Both (1.11) and (1.12) are equivalent and essentially state that lines of electrostatic field do not form closed loops. This as we shall see is not true for the general time-dependent electric field.

A line or surface of constant ϕ is called an equipotential. If $\mathrm{d}\boldsymbol{r}$ is a displacement on an equipotential surface, $\mathrm{d}\phi = \mathrm{d}\boldsymbol{r}.\boldsymbol{\nabla}\phi = 0$, and so $\boldsymbol{\nabla}\phi$ and $\boldsymbol{E}$ are perpendicular to equipotentials.

Gauss's theorem

If S is a closed surface with positive outward normal and $\boldsymbol{r}$ is the radius vector from the origin to an element $\mathrm{d}\boldsymbol{S}$ of S the scalar product $\boldsymbol{r}.\mathrm{d}\boldsymbol{S}/r$ is the projection of $\mathrm{d}\boldsymbol{S}$ on a sphere of radius r, and the product $\boldsymbol{r}.\mathrm{d}\boldsymbol{S}/r^3$ is the projection on a unit sphere, about the origin. It follows that the closed surface integral

$$\oint_S \frac{\boldsymbol{r}.\mathrm{d}\boldsymbol{S}}{r^3}$$

is equal to 4π if S encloses the origin. If however S does not enclose the origin, $\boldsymbol{r}$ cuts S at two points and contributions from these two intersections cancel. The integral is then zero. The electric field at $\boldsymbol{r}$ due to a charge q at the origin is $q\boldsymbol{r}/4\pi\epsilon_0 r^3$. It now follows that the flux of electric field lines out of any closed surface S i.e. the integral $\oint \boldsymbol{E}.\mathrm{d}\boldsymbol{S}$ is zero if S does not contain q and q/ϵ_0 if it does. In general if ρ is the volume charge density and S the surface of the volume V we have

$$\oint_S \boldsymbol{E}.\mathrm{d}\boldsymbol{S} = \frac{1}{\epsilon_0}\int \rho\,\mathrm{d}V. \qquad (1.13)$$

This is known as Gauss's theorem and, by taking an infinitesimal volume, we can convert it to the differential equation

$$\nabla.\boldsymbol{E} = \rho/\epsilon_0. \qquad (1.14)$$

These two equivalent equations state that lines of $\boldsymbol{E}$ begin on positive charge and end on negative charge.

Conductors

Conductors contain mobile charge and, under the influence of an electric field, this charge moves until it sets up a field which just cancels the original field in the conductor. It will, of course, alter the field outside the conductor as well but it need not reduce it to zero except in the conductor. If $\boldsymbol{E}$ is zero in a conductor $\nabla\phi$ is zero and ϕ is constant. Thus, in electrostatics conductors are equipotentials, and lines of $\boldsymbol{E}$ meet conducting surfaces at right angles. The charge density $\rho = \epsilon\nabla.\boldsymbol{E}$ is also zero and any charge on a conductor can only reside on the surface. In Fig. 1.3 we show a metal surface and the field E_0 outside the surface. If we apply Gauss's theorem to the 'pill-box'

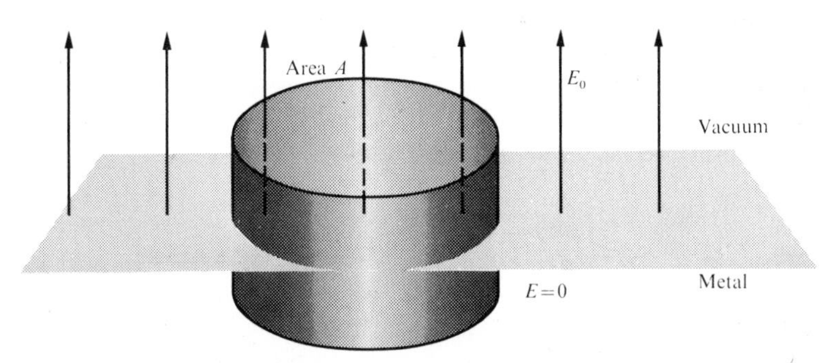

FIG. 1.3. The field at a metal surface.

volume we find that the charge in this volume is $\epsilon_0 E_0 A$. There is therefore a surface charge on the metal equal to $\epsilon_0 E_0$ per unit area. This charge resides in a thin layer at the surface. In metals there are some 10^{29} electrons (each of charge $-1{\cdot}6\times10^{-19}$ coulomb) per cubic metre. The mobile charge density is therefore about 10^{10} coulomb m^{-3}. Since $\epsilon_0 \sim 10^{-11}$ farad m^{-1}, if $\boldsymbol{E} = 10^8$ V m^{-1} (a very large field) the surface charge is only 10^{-3} coulomb

m^{-2} and can be supplied from a surface layer only 10^{-13} m thick i.e. less than one atomic layer. In semiconductors on the other hand the mobile charge density is less by a factor of perhaps 10^{-8} and it is possible to remove all the mobile charge from a thin wafer of material. This is used in the field-effect transistor.

In the surface layer the force per unit volume acting on the conductor is $\rho E = \epsilon_0 E(\partial E/\partial x)$ and the force per unit area is obtained by integrating this through the layer from a region where $E = 0$ to the surface where $E = E_0$. The result is $\frac{1}{2}\epsilon_0 E_0^2$. Since $\sigma = \epsilon_0 E_0$ is the charge per unit area the force per unit area is $\frac{1}{2}\sigma E_0$ not, be it noted, σE_0. The average field within the charge layer is $\frac{1}{2}E_0$ not E_0.

If two parallel conducting plates of area A are placed a distance x apart, as shown in Fig. 1.4, and the upper plate is given a charge $+q$, a charge $-q$ is

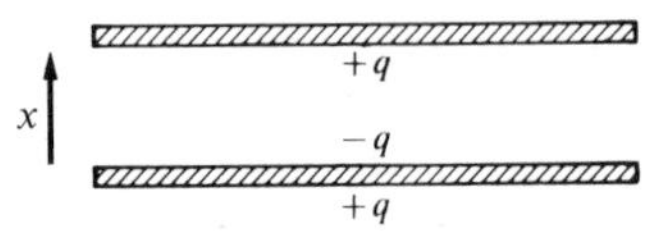

FIG. 1.4. A charged metal plate near an uncharged plate.

induced in the near face of the lower plate leaving $+q$ on the outer face. If the lower plate is earthed $+q$ is transferred to ground and the lower plate has a net charge $-q$. We can break the earth connection and transfer this charge to a third body by moving the lower plate away. Repetition of this process n times transfers $-nq$ to the third body. Induction effects such as this are used in many high-voltage electrostatic machines. With the lower plate earthed the field between the plates obtained from Gauss's theorem is $-q/\epsilon_0 A$ and so the potential difference is $\phi = qx/\epsilon_0 A$. The capacitance of the system is defined as the ratio q/ϕ, and is therefore $C = \epsilon_0 A/x$. In general the capacitance between any two conductors can be expressed as

$$C = \frac{\epsilon_0}{\phi}\int_S E_n \, \mathrm{d}S, \tag{1.15}$$

where the integral is over the surface of the positive plate and E_n is the field away from the plate, or alternatively the integral is over the negative plate with E_n directed into the plate. The work done to add a charge $\mathrm{d}q$ to a capacitor already at a potential ϕ is $\phi\, \mathrm{d}q = q\, \mathrm{d}q/C$ and so the work required to charge it is $q^2/2C = \frac{1}{2}C\phi^2 = \frac{1}{2}q\phi$. We can express this in terms of the electric field E in the neighborhood of the plates (see Chapter 6) as the volume integral

$$U = \int_V \tfrac{1}{2}\epsilon_0 E^2 \, \mathrm{d}V. \tag{1.16}$$

In this sense an electric field stores energy with a density $\frac{1}{2}\epsilon_0 E^2$.

Properties of the potential

If $\phi(\boldsymbol{r})$ is known as a function of $\boldsymbol{r}$ it gives $\boldsymbol{E}$ as $-\boldsymbol{\nabla}\phi$ and so the primary problem in solving electrostatic problems is to obtain ϕ. Since $\boldsymbol{\nabla}.\boldsymbol{E} = \rho/\epsilon_0$, ϕ satisfies Poisson's equation,

$$\nabla^2\phi = -\rho/\epsilon_0, \tag{1.17}$$

where

$$\nabla^2\phi = \boldsymbol{\nabla}.(\boldsymbol{\nabla}\phi) = \frac{\partial^2\phi}{\partial^2 x}+\frac{\partial^2\phi}{\partial y^2}+\frac{\partial^2\phi}{\partial z^2}, \tag{1.18a}$$

or, in spherical polar coordinates,

$$\nabla^2\phi = \left(\frac{1}{r^2}\frac{\partial}{\partial r}r^2\frac{\partial}{\partial r}+\frac{1}{r^2\sin\theta}\frac{\partial}{\partial\theta}\sin\theta\frac{\partial}{\partial\theta}+\frac{1}{r^2\sin^2\theta}\frac{\partial^2}{\partial\varphi^2}\right)\phi. \tag{1.18b}$$

If ρ is known everywhere we can solve this equation immediately for, if in eqn (1.8) we replace the individual charges by a density ρ, we have

$$\phi(\boldsymbol{r}(1)) = \int_V \frac{\rho(\boldsymbol{r}(2))\,\mathrm{d}V(\boldsymbol{r}(2))}{4\pi\epsilon_0 r(1,2)}. \tag{1.19}$$

However ρ must be known everywhere, including the surface of conductors, and so this is not usually a practical proposition.

We shall give an important example of the use of (1.19) in Chapter 6. More usually we have to find ϕ either when the potentials on a set of conductors are given or when the positions of certain charges are given and there are also electrodes present. In the first case we are interested in the value of ϕ in the empty space between the electrodes. Here ϕ satisfies Laplace's equation

$$\nabla^2\phi = 0. \tag{1.20}$$

A general theorem assures us that there is only one solution of this equation which fits any prescribed set of boundary values i.e. the values of ϕ on the electrodes. We can therefore proceed to solve (1.20) either by a process of approximations or by organized guesswork, confident that any solution we obtain, which satisfies this equation will be correct. We give two examples to illustrate these methods.

Suppose that we have a set of wires, parallel to the y-axis in the plane $z = 0$ and equally spaced at $x = \pm n(a/2)$ where $n = 0, 1, 2$ etc. Wires at positions where n is even are at a positive potential ϕ_0, wires at odd positions are at $-\phi_0$. It is obvious that ϕ is independent of y and varies periodically with x. Thus, as far as the x-dependence goes, ϕ is of the general form

$$\phi = \sum_{k=-\infty}^{\infty} a_k(z)\cos\frac{2\pi kx}{a}.$$

This will only satisfy Laplace's equation if $a_k(z)$ is of the form

$$a_k(z) = A_k \exp(2\pi kz/a)$$

where A_k is a constant. For $z > 0$ we can only use negative values of k and for $z < 0$ positive values. Thus for $z > 0$ the potential is

$$\phi(x, z) = \sum_{k=1}^{\infty} A_k \exp\frac{-2\pi kz}{a} \cos\frac{2\pi kx}{a}. \tag{1.21}$$

At $z = 0$, ϕ must fit the prescribed potential and this determines the values of the coefficients A_k. The coefficient A_1 is easily shown to be $(2/\pi)\phi_0$. Thus at a distance $z \gg a$ from the grid, where all the other terms are small,

$$\phi(x, z) \sim (2/\pi)\phi_0 \exp(-2\pi z/a)\cos(2\pi x/a).$$

As a second example we consider the potential $\phi = Axy$ which clearly satisfies Laplace's equation. The equipotentials are hyperbolic sheets and this is therefore the potential near the origin due to the hyperbolic electrodes shown in Fig. 1.5. When both prescribed charges and electrodes are present the

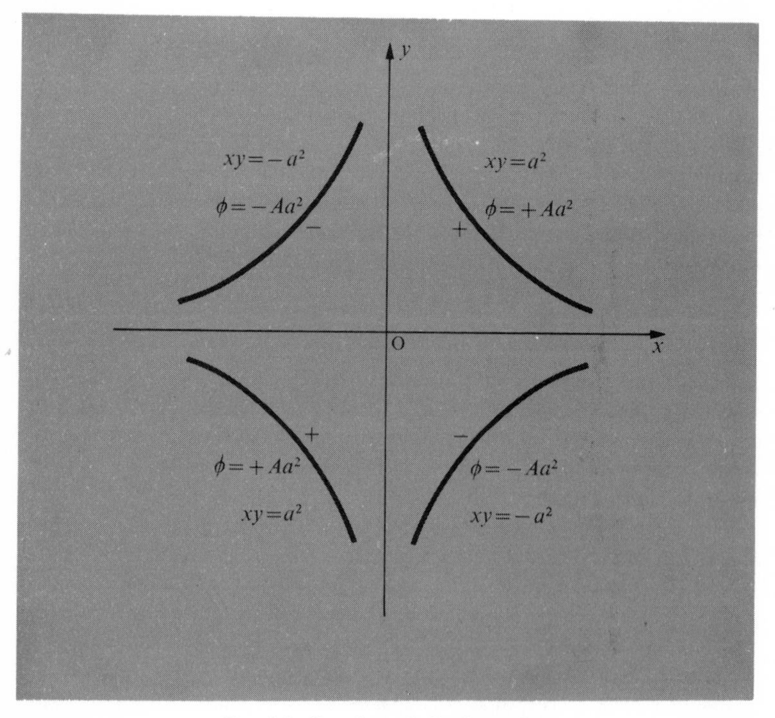

FIG. 1.5. Four hyperbolic electrodes.

problem is more difficult. There is a formal procedure based on the use of Green's functions for solving this type of problem. It is not of much practical use. We normally resort to guesswork based on past experience of similar problems. The method of images described in the next example is a typical procedure.

The potential due to point charges q at $(0, 0, l)$ and $-q$ at $(0, 0, -l)$ is

$$\phi = \frac{q}{4\pi\epsilon_0}\left\{\frac{1}{[x^2+y^2+(z-l)^2]^{\frac{1}{2}}} - \frac{1}{[x^2+y^2+(z+l)^2]^{\frac{1}{2}}}\right\} \tag{1.22}$$

and vanishes on the plane $z = 0$. This is therefore for $z > 0$, also the potential due to a charge q at $(0, 0, l)$ in front of a conducting plane in the xy plane at $z = 0$.

A point charge q placed in front of a conducting plate induces a net charge $-q$ on the near surface of the plate. We are, however, especially in studies of vacuum electronics, plasma physics, and semiconductor devices, often interested in the charges induced on two plates by an electron of charge $q = -e$ between the plates. We can solve this problem as follows.

If two plates are maintained at potentials $\phi = 0$ and $\phi = \phi_0$ by a battery and a charge q is taken from the plate at $\phi = 0$ to a position where the potential is ϕ the mechanical work required is $q\phi$. This is delivered to the battery. The induced charge on the plate at ϕ_0 must therefore be given by $-q(\phi/\phi_0)$. The ratio ϕ/ϕ_0 depends only on the geometry of the plates not on ϕ_0 the battery voltage, and so if we have two plates connected by a wire we can still use this relation to calculate the induced charges.

If a curve satisfies

$$\frac{d^2y}{dx^2} = 0$$

it is straight i.e. has no curvature. If a surface of height z satisfies

$$\frac{d^2z}{dx^2}+\frac{d^2z}{dy^2} = 0$$

the sum of the two principal curvatures at each point is zero. It has saddle-points and valleys but no peaks or troughs. Laplace's equation

$$\frac{\partial^2\phi}{\partial x^2}+\frac{\partial^2\phi}{\partial y^2}+\frac{\partial^2\phi}{\partial z^2} = 0$$

has a similar meaning. The electrostatic potential has no absolute maxima or minima in empty space. This is known as Earnshaw's theorem and shows that a charged particle e.g. an electron, can never be at rest under the influence of electrostatic forces alone.

PROBLEMS

1.1. Is it possible to adopt a convention in which like charges attract? Consider three bodies with charges $+q$, $+q$, and $-q$.

1.2. Show that Kirchhoff's circuit law in a.c. circuits—the sum of all the currents leading to a node is zero—is related to eqn (1.4b).

1.3. Show that

$$\frac{1}{|\boldsymbol{r}+\boldsymbol{l}|} \sim \frac{1}{r}+\boldsymbol{l}.\boldsymbol{\nabla}\left(\frac{1}{r}\right) \quad \text{if} \quad r \gg l.$$

1.4. Point charges $-q$ and $+q$ are placed at the origin and the point $\boldsymbol{l}$; show that the potential at distant points $r \gg l$ can be expressed as

$$\phi = \frac{q\boldsymbol{l}.\boldsymbol{r}}{4\pi\epsilon_0 r^3}.$$

1.5. Prove that $\boldsymbol{\nabla} \wedge (\boldsymbol{\nabla}\phi) = 0$. Compare this result with eqn (1.12). We shall see later that the general electromagnetic field $\boldsymbol{E}$ does not satisfy $\boldsymbol{\nabla} \wedge \boldsymbol{E} = 0$. Can this field be derived from a scalar potential?

1.6. Prove that in an incompressible fluid the conservation of mass implies that the velocity vector $\boldsymbol{v}$ satisfies $\boldsymbol{\nabla}.\boldsymbol{v} = 0$.

1.7. Use eqn (1.15) to derive the capacitance between two concentric spherical shells of radii a and $b > a$, and the capacitance per unit length of two coaxial cylinders.

1.8. In problem 1.4 take $\boldsymbol{l}$ to be along the polar axis of spherical coordinates and express ϕ in spherical coordinates. Verify that the potential satisfies Laplace's equation. If $\phi = Ar\cos\theta$ what is the corresponding electric field?

1.9. Show that $\phi = xyz$ satisfies Laplace's equation. Express xyz in spherical polar coordinates with the polar axis along the z-axis and verify that it satisfies Laplace's equation in these coordinates. Verify also that xyz/r^7 is a solution.

1.10. Verify that $\cos\alpha x \cos\beta y \exp(-\gamma z)$ is a solution of Laplace's equation if $\alpha^2+\beta^2 = \gamma^2$.

1.11. What is the force on a particle of charge q distant l from a conducting plane?

1.12. Electrons of charge q move with uniform normal component of velocity v between two parallel conducting planes a distance d apart connected by an external wire. Show that each electron produces a current pulse of amplitude qv/d, lasting a time d/v in the external wire.

1.13. A screened room is constructed by embedding chicken-wire of 1 in. mesh in the walls. Use the result of eqn (1.21) to explain qualitatively why external electrostatic interference will be negligible at distances of more than a few inches from the walls.

1.14. A coin of radius r and thickness $t \ll r$ is placed in a uniform electrostatic field $\boldsymbol{E}$ due to distant charges with its axis parallel to $\boldsymbol{E}$. What are the charges induced on its plane faces?

1.15. A body of charge q is placed inside an isolated, closed metal vessel without touching it. Apply Gauss's theorem to a surface entirely within the metal and show that the charge on the outer surface of the vessel is q. A metal tube is connected to one plate of a capacitor (C) with its other plate earthed and a particle of charge q is projected through the tube. How does the voltage across C vary as the particle enters and passes through the tube?

1.16. The potential in a vacuum tube varies with the distance from the cathode as $\phi = Ax^{\frac{4}{3}}$. How does the space-charge density vary? What is the current density in the electron beam? (Electrons of energy 2 500 electronvolts travel with a velocity of $3 \cdot 10^7$ m s^{-1}.)

1.17. Can a particle of mass m and charge q be in stable equilibrium under the influence of electrostatic and gravitational fields alone?

1.18. Estimate the electrostatic capacitance of the human body. Use $\epsilon_0 \sim 10^{-11}$ farad m^{-1}. Compare the result with the capacitance of a radio tuning capacitor.

1.19. Above the plane $z = 0$ the electrostatic potential is $\phi = \cos \alpha x \exp(-\alpha z)$ and below the plane it is $\phi = \cos \alpha x \exp(\alpha z)$. Find the surface charge distribution in the plane $z = 0$.

1.20. There is a uniform charge density ρ independent of x and y in the laminar region $-a < z < 0$ and a similar charge density $-\rho$ in the region $0 < z < a$. Show that the electric fields in the two regions $z < -a$ and $z > a$ are equal. If both the field and the potential are zero for $z < -a$ what is the potential in the region $z > a$?

1.21. The electron beam in a cathode ray tube passes between two *short* plane deflection plates. Show that the deflection for a given charge on the plates is independent of the separation of the plates. (Do *not* neglect edge effects.)

2. Magnetism

The magnetic field

ALTHOUGH magnetic effects occur in the absence of measurable currents we now know that all magnetism is associated with currents, either on a laboratory scale, or on an atomic scale. Even the magnetic moment of the neutron is to be regarded as evidence that the neutron has electromagnetic structure. The force acting on a current I flowing in an element $\mathrm{d}\boldsymbol{l}$ of a wire near a magnet is at right angles to $\mathrm{d}\boldsymbol{l}$. If $\mathrm{d}\boldsymbol{F}$ is the force on an element $\mathrm{d}\boldsymbol{l}$ at $\boldsymbol{r}$ the magnetic field $\boldsymbol{B}(\boldsymbol{r})$ is defined so that

$$\mathrm{d}\boldsymbol{F} = I\,\mathrm{d}\boldsymbol{l} \wedge \boldsymbol{B}(\boldsymbol{r}) \tag{2.1}$$

and the force per unit volume acting in a region where there is a current density $\boldsymbol{J}$ is

$$\boldsymbol{f} = \boldsymbol{J} \wedge \boldsymbol{B}(\boldsymbol{r}). \tag{2.2}$$

If a volume contains a particle of velocity $\boldsymbol{v}$ and charge q the volume integral of $\boldsymbol{J}$ over a region around the particle is $\int \boldsymbol{J}\,\mathrm{d}V = q\boldsymbol{v}$ and the force on the particle is

$$\boldsymbol{F} = q\boldsymbol{v} \wedge \boldsymbol{B}. \tag{2.3}$$

In this book we refer to the vector $\boldsymbol{B}$ as the magnetic field; this is consistent with our general notion that a field is defined in terms of the force it exerts, but many books call $\boldsymbol{B}$ either the magnetic induction or flux density.

The field at $\boldsymbol{r}(1)$ due to a current element $I\,\mathrm{d}\boldsymbol{l}(\boldsymbol{r}(2))$ at $\boldsymbol{r}(2)$ is at right angles to $\mathrm{d}\boldsymbol{l}$ and $\boldsymbol{r}(1, 2)$ and obeys an inverse square law. This is all expressed by the relation

$$\mathrm{d}\boldsymbol{B}(\boldsymbol{r}(1)) = \frac{\mu_0}{4\pi}\cdot\frac{I\,\mathrm{d}\boldsymbol{l}(\boldsymbol{r}(2)) \wedge \boldsymbol{r}(1, 2)}{r^3(1, 2)} \tag{2.4a}$$

where μ_0 is chosen to define the units of I. In terms of a current density $\boldsymbol{J}$ this equation becomes

$$\mathrm{d}\boldsymbol{B}(\boldsymbol{r}(1)) = \mu_0\frac{\boldsymbol{J}(\boldsymbol{r}(2)) \wedge \boldsymbol{r}(1, 2)\,\mathrm{d}V(\boldsymbol{r}(2))}{4\pi r^3(1, 2)}. \tag{2.4b}$$

When this result is combined with eqn (2.1) the equation is known as the Biot–Savart Law. The field at $\boldsymbol{r}(1)$ due to a complete current distribution is

$$\boldsymbol{B}(\boldsymbol{r}(1)) = \int_{V_2} \frac{\mu_0\boldsymbol{J}(\boldsymbol{r}(2)) \wedge \boldsymbol{r}(1, 2)\,\mathrm{d}V(\boldsymbol{r}(2))}{4\pi r^3(1, 2)}, \tag{2.5}$$

and, due to a complete circuit characterized by a line element $\mathrm{d}\boldsymbol{l}(\boldsymbol{r}(2))$,

$$\boldsymbol{B}(\boldsymbol{r}(1)) = \oint \frac{\mu_0 I_2\,\mathrm{d}\boldsymbol{l}(\boldsymbol{r}(2)) \wedge \boldsymbol{r}(1, 2)}{4\pi r^3(1, 2)}. \tag{2.6}$$

The force due to this field acting on another complete circuit with a current I^1 is

$$F_{12} = \oint I_1 \, dl(r_1) \wedge B(r(1)). \tag{2.7}$$

When these two results are combined and rearranged (which is rather tedious) the force can be expressed as

$$F_{12} = +\frac{\mu_0}{4\pi} I_1 I_2 \nabla_1 \oint_1 \oint_2 \frac{dl(r(1)) . dl(r(2))}{r(1, 2)}. \tag{2.8}$$

Since ∇_2 (with respect to $r(2)$) of the integral is $-\nabla_1$ of the integral, action and reaction balance and $F_{12} = -F_{21}$.

Units

If two circuits are wired in series, so that they carry the same current I, the force between the circuits,

$$F_{12} = +\frac{\mu_0}{4\pi} I^2 \nabla_1 \oint_1 \oint_2 \frac{dl(r(1)) . dl(r(2))}{r(1, 2)}, \tag{2.9}$$

involves the constant μ_0 and a geometrical integral which can be evaluated exactly when the dimensions of the circuits are known. Thus, if lengths are expressed in metres and the force in newtons, once we have fixed the value of μ_0, the unit of I is fixed. This *current balance* can also be used to establish an accurate experimental standard of current. The constant μ_0 is arbitrarily given the exact value $\mu_0 = 4\pi \times 10^{-7}$ (henry m^{-1}) and then I is measured in amperes (A). The international standard ampere is calibrated at the National Bureau of Standards in Washington. (The ampere is the primary electric unit. The coulomb, or unit of charge, is the ampere second.)

As we shall later see the velocity of light in vacuum is $c = (\mu_0 \epsilon_0)^{-\frac{1}{2}}$, and since c can be measured with great precision the constant ϵ_0 is defined as $1/\mu_0 c^2 = 10^7/4\pi c^2$ where c is the experimental value of the velocity of light. The units of ϵ_0 are farad m^{-1}.

The secondary electrical standard is the ohm (Ω) or unit of resistance. This is calibrated by comparing a physical resistance in an a.c. bridge (see e.g. G. Lancaster *D.C. and A.C. Circuits* Oxford Physics Series) with a capacitor whose capacitance has been calculated from its geometry using the value $\epsilon_0 = 10^7/4\pi c^2$. The unit of potential, the volt V is the ohm ampere, the potential drop across 1 ohm when 1 ampere flows. The unit of power, the watt, is one volt ampere. Electrical standards are so precise and measurements of power and energy so imprecise that for all practical purposes the watt second is equal to the joule which is twice the kinetic energy of a standard kilogram travelling at 1 m s^{-1}. In most laboratories the practical standards are a standard potential from a standard cell and a standard resistance. The laboratory ampere then becomes the derived unit. Only in the most precise work is this difference significant.

Properties of the field $\boldsymbol{B}$

The field $\boldsymbol{B}$ at $\boldsymbol{r}(1)$ due to a current distribution is given by eqn (2.5) which we can express, using (1.5) as

$$\boldsymbol{B}(\boldsymbol{r}(1)) = -\int_{V_2} \frac{\mu_0}{4\pi} \boldsymbol{J}(\boldsymbol{r}(2)) \wedge \nabla_1\left(\frac{1}{r(1,2)}\right) \mathrm{d}V(\boldsymbol{r}(2)).$$

Because

$$\nabla_1 \wedge \left(\frac{\boldsymbol{J}(\boldsymbol{r}(2))}{r(1,2)}\right) = -\boldsymbol{J} \wedge \nabla_1\left(\frac{1}{r(1,2)}\right) + \frac{1}{r(1,2)} \nabla_1 \wedge \boldsymbol{J}(\boldsymbol{r}(2)) \tag{2.10}$$

and $\nabla_1 \wedge \boldsymbol{J}(\boldsymbol{r}(2)) = 0$ (because $\boldsymbol{J}$ is not a function of $\boldsymbol{r}(1)$) we have

$$\boldsymbol{B}(\boldsymbol{r}(1)) = \nabla_1 \wedge \int \frac{\mu_0 \boldsymbol{J}(\boldsymbol{r}(2))\, \mathrm{d}\nabla(\boldsymbol{r}(2))}{4\pi r(1,2)}. \tag{2.11}$$

According to (2.11) the vector $\boldsymbol{B}$ can be expressed as

$$\boldsymbol{B}(\boldsymbol{r}(1)) = \nabla_1 \wedge \boldsymbol{A}(\boldsymbol{r}(1)) \tag{2.12}$$

where the *vector potential* $\boldsymbol{A}$ is

$$\boldsymbol{A}(\boldsymbol{r}(1)) = \int \frac{\mu_0 \boldsymbol{J}(\boldsymbol{r}(2))\, \mathrm{d}V(\boldsymbol{r}(2))}{4\pi r(1,2)}. \tag{2.13}$$

and, since the divergence of the curl of a vector is identically zero, we therefore have

$$\nabla . \boldsymbol{B} = 0. \tag{2.14}$$

Lines of $\boldsymbol{B}$ do not end. As many lines of $\boldsymbol{B}$ leave a volume as enter it. Lines of $\boldsymbol{B}$ behave like streamlines in an incompressible fluid. It is more difficult to obtain the value of $\nabla \wedge \boldsymbol{B}$ from eqn (2.5) although it is still a straightforward, if tedious, piece of mathematics. The result, however, is simple

$$\nabla \wedge \boldsymbol{B} = \mu_0 \boldsymbol{J}. \tag{2.15}$$

This is usually obtained by a rather misleading argument as follows. According to eqn (2.6) the field $\boldsymbol{B}$ near a long straight wire forms circular lines about the wire. If the current in the wire is I the magnitude of the field at a radial distance r is $\mu_0 I/2\pi r$. The line integral around a field line is therefore $\mu_0 I$. Thus we have

$$\oint \boldsymbol{B}.\mathrm{d}r = \mu_0 I, \tag{2.16}$$

where I is the current linking the path of integration. From this and the definition of the curl ($\nabla \wedge \boldsymbol{B}$) we obtain (2.15). The argument is misleading because it does not pay proper attention to the conditions at the ends of the wire.

The properties of the field $\boldsymbol{B}$ are very different from those of the electrostatic field $\boldsymbol{E}$. Lines of the static electric field do not form closed loops and end on charges. Lines of $\boldsymbol{B}$ form closed loops and never end.

Equation (2.6) affords the easiest way of calculating $\boldsymbol{B}$ due to currents in wires but, when we have to deal with distributed currents, e.g. in atoms, eqns (2.12) and (2.13) are generally simpler to use. Unfortunately, unlike the scalar electrostatic potential, the vector magnetic potential $\boldsymbol{A}$ has no simple physical meaning.

The Lorentz force

The force on a particle of charge q and velocity $\boldsymbol{v}$ moving through fields $\boldsymbol{E}$ and $\boldsymbol{B}$ is

$$\boldsymbol{F} = q\boldsymbol{E} + q\boldsymbol{v} \wedge \boldsymbol{B}, \tag{2.17}$$

and the force per unit volume in a region where the charge density is ρ and the current density $\boldsymbol{J}$ is

$$\boldsymbol{f} = \rho\boldsymbol{E} + \boldsymbol{J} \wedge \boldsymbol{B}. \tag{2.18}$$

Consider a straight conductor of length l moving with a velocity $\boldsymbol{v}$ perpendicular to its length and a magnetic field $\boldsymbol{B}$ as shown in Fig. 2.1. Mobile charges in

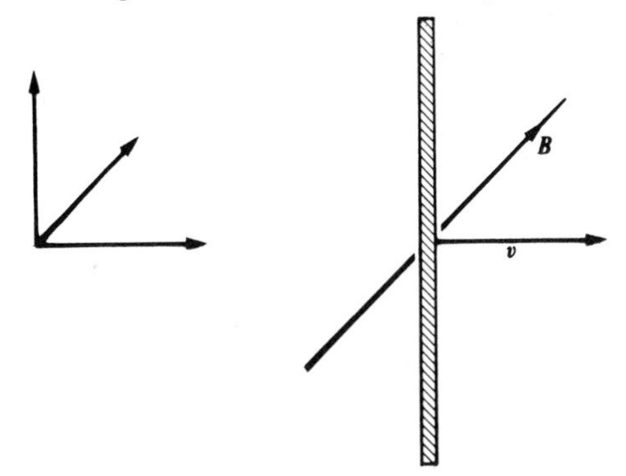

FIG. 2.1. A moving conductor.

the conductor will experience a force driving them along the wire and a positive charge will accumulate at one end of the wire and a negative charge at the other until an electric field $\boldsymbol{E}$ is set up which brings them to rest. From eqn (2.17) we see that this field will be $\boldsymbol{E} = -\boldsymbol{v} \wedge \boldsymbol{B}$ and the potential difference between the ends of the wire will be $|l\boldsymbol{v} \wedge \boldsymbol{B}|$. If, as in Fig. 2.2, the ends of the wire make contact with two conducting rails the e.m.f. electromotive

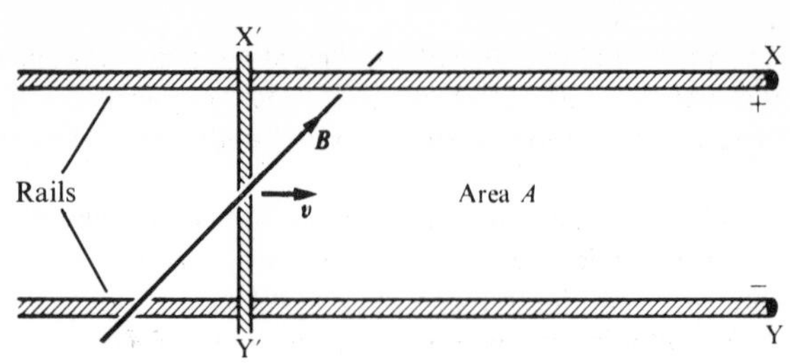

FIG. 2.2. A wire moving along two conducting rails.

force) or potential difference between X and Y will be equal to the rate of decrease of the flux BA linking the circuit XX′Y′Y.

Summary

The two equations $\nabla.\boldsymbol{B} = 0$ and $\nabla \wedge \boldsymbol{B} = \mu_0 \boldsymbol{J}$ which express the primary properties of the magnetic field $\boldsymbol{B}$ are direct mathematical consequences of the Biot–Savart law. The Lorentz force-equation is essentially a statement about the meaning of the fields $\boldsymbol{B}$ and $\boldsymbol{E}$. The equation $\nabla.\boldsymbol{B} = 0$ like the equation $\nabla.\boldsymbol{E} = \rho/\epsilon_0$ is one of Maxwell's field equations and is preserved in a general treatment of time-dependent fields. The derivation of the equation $\nabla \wedge \boldsymbol{B} = \mu_0 \boldsymbol{J}$, either by purely mathematical steps, or by considering a long straight wire, ultimately depends, however, on the assumption that the currents flow in closed circuits or equivalently that $\nabla.\boldsymbol{J}$ and therefore $\dot{\rho}$ are zero. If $\nabla.\boldsymbol{J} \neq 0$ we cannot deduce anything about $\nabla \wedge \boldsymbol{B}$ whatsoever. In Chapter 4 we shall see how Maxwell, by an inspired hypothesis, obtained the correct generalization of the expression for $\nabla \wedge \boldsymbol{B}$.

The reader may feel that our treatment of $\boldsymbol{B}$ has been rather formal and less easy to follow than the conventional treatment using simple examples. In the author's view it is not possible to introduce the properties of $\boldsymbol{B}$, except rather formally, without giving rise to misunderstanding. In particular the use of magnetic double layers or magnetic poles tends to obscure the fundamental properties of the field.

PROBLEMS

2.1. A field $\boldsymbol{B}$ is applied parallel to a diameter of a circular loop of wire in which a current I is flowing. What is the force on the loop as a whole and what is the magnitude and direction of the couple, or torque, acting on the loop?

2.2. An electron of velocity $\boldsymbol{v}$ moves in a uniform magnetic field $\boldsymbol{B}$, show that unless $\boldsymbol{v}$ is parallel to $\boldsymbol{B}$ the electron trajectory is a helix.

2.3. Show that the field $\boldsymbol{B}$ near a long straight wire carrying a current I forms circular lines of force around the wire and that the magnitude of the field at a radius r is $\mu_0 I/2\pi r$.

2.4. Calculate the field at the centre of a circular loop of wire of radius r carrying a current I.

2.5. A long helix, or solenoid, has n turns per metre and carries a current I. What is the field on the axis of the solenoid?

2.6. The vector potential is given as $\boldsymbol{A} = (y, -x, 0)$. What is the direction of the magnetic field derived from $\boldsymbol{A}$ and its magnitude?

2.7. An electron moves through a region where there is a steady magnetic field $\boldsymbol{B}$. Show that its kinetic energy is constant.

2.8. A copper rod of 1-metre length moves with a velocity of 1 metre s^{-1} normal to its length and normal to the earth's magnetic field (about 10^{-4} tesla). What is the e.m.f. developed between its ends? (The tesla T is the SI unit of field. The c.g.s. unit, the gauss, is 10^{-4} T.)

2.9. Two thin beams of electrons each of velocity v and carrying a current I move parallel to each other at a separation r. Find the total (electrostatic and magnetic) force acting, per unit length between the beams. What happens as $v \to c = (\mu_0 \epsilon_0)^{-\frac{1}{2}}$?

2.10. An electron of charge $q = -e$ and mass m initially at rest at the origin is acted on by an electric field with the single component $E_x = E_0 \cos \omega t = R_e E_0 \exp(j\omega t)$, in the presence of a uniform static magnetic field B parallel to the z-axis. Discuss the motion of the electron paying particular attention to the case when

$$\omega = \omega_c = \frac{e}{m} B.$$

You may find it useful to introduce the variables $r_+ = x + jy$ and $r_- = x - jy$.

2.11. In a region free of current the direction of the magnetic field is found to be constant. Show that its magnitude must also be constant.

3. Electromagnetic induction

Moving conductors

WHENEVER a conductor moves in a magnetic field the Lorentz force on the mobile carriers in the conductor leads to charge separation and an induced e.m.f. If $\mathrm{d}\boldsymbol{l}$ is a line element in a conductor moving with velocity $\boldsymbol{v}$ in a field $\boldsymbol{B}$ the e.m.f. produced between the ends of $\mathrm{d}\boldsymbol{l}$ is $-\mathrm{d}\boldsymbol{l}(\boldsymbol{v} \wedge \boldsymbol{B})$. This effect is seen at its simplest in the homopolar generator in which a copper disc rotates about its axis which is parallel to a uniform field $\boldsymbol{B}$. A radial segment $\mathrm{d}\boldsymbol{r}$ at $\boldsymbol{r}$ moves with a velocity ωr where ω is the angular velocity and the induced e.m.f. is radial and equal to $\omega r B\,\mathrm{d}r$. The total e.m.f. between the axis and the periphery at radius a is $\frac{1}{2}\omega a^2 B$. It can be picked up by contacts as shown in Fig. 3.1 and used to drive a current in a load R.

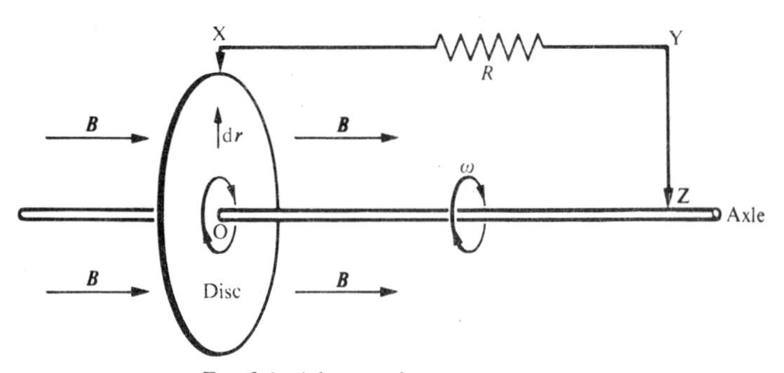

FIG. 3.1. A homopolar generator.

If a closed loop of wire moves, or changes shape, in a steady field $\boldsymbol{B}$ the total e.m.f. induced in the loop is the line integral

$$V = +\oint(\boldsymbol{v} \wedge \boldsymbol{B}).\mathrm{d}\boldsymbol{l} = +\oint \boldsymbol{B}.(\mathrm{d}\boldsymbol{l} \wedge \boldsymbol{v}). \tag{3.1}$$

Now $\mathrm{d}\boldsymbol{l} \wedge \boldsymbol{v}$ is the rate at which the line element is sweeping out an area and $-\boldsymbol{B}.(\mathrm{d}\boldsymbol{l} \wedge v)$ is the rate at which new lines of $\boldsymbol{B}$ are included within the area of the loop. The total number of lines of $\boldsymbol{B}$ linking the loop is known as the linked magnetic flux Ψ and so the integral in (3.1) is $\dot{\Psi}$ and the e.m.f. is

$$V = -\dot{\Psi}. \tag{3.2}$$

Because $\nabla.\boldsymbol{B} = 0$ the total number of lines of $\boldsymbol{B}$ crossing any open area S bounded by a given line, in this case the loop of wire, is the same, and so we can express Ψ as

$$\Psi = \int_S B_n \, \mathrm{d}S = \int_S \boldsymbol{B} . \mathrm{d}\mathbf{S}. \tag{3.3}$$

For this reason $\boldsymbol{B}$ is sometimes called the magnetic flux density.

Eqn (3.2) refers only to a continuous loop of wire, although the loop may change shape during its motion. It does not apply to a system such as the homopolar generator in Fig. 3.1, where new conducting elements are continually entering the loop OXYZO as the disc rotates. Indeed the flux linking this loop is not only constant, it is zero.

The flip-coil fluxmeter

If a small plane coil of area A and N turns is placed with its plane normal to the field $\boldsymbol{B}$ of a magnet, the flux linking the coil is NAB and, as the coil is withdrawn from the magnet, the induced e.m.f. is

$$-\frac{\mathrm{d}}{\mathrm{d}t}(NAB)$$

so that, if the coil is connected to a galvanometer of resistance R, the total charge registered by the galvanometer, as the coil is withdrawn completely from the field, is

$$q = -\frac{1}{R} \int_0^\infty \frac{\mathrm{d}}{\mathrm{d}t}(NAB) \, \mathrm{d}t = \frac{NAB}{R}. \tag{3.4}$$

This gives us a method of measuring fields directly in terms of the Lorentz force that they exert on charge carriers.

Faraday's law

If a flip-coil fluxmeter is set up and, instead of removing the coil from the magnet we remove the magnet from the vicinity of the coil we also find that there is an induced e.m.f. and it is again given by (3.4). This result is intuitively reasonable since we might have expected that only the relative displacement of the coil and magnet would be significant. The discovery by Faraday that an e.m.f. was induced in a stationary coil when the flux linking it changed has however profound consequences. We cannot explain this effect in terms of the Lorentz force on moving carriers, for in this case the carriers do not move, and we are therefore forced to formulate a new law. If $\mathrm{d}\boldsymbol{l}$ is a line element of a closed loop of wire, fixed in space in a region where the magnetic field $\boldsymbol{B}$ is changing, then the e.m.f. induced in the loop is again $-\dot{\Psi}$ and we can express this as

$$\oint \boldsymbol{E}.\mathrm{d}\boldsymbol{l} = -\int_S \dot{\boldsymbol{B}}.\mathrm{d}\boldsymbol{S}. \tag{3.5}$$

The physical presence of the wire plays no part in this effect and eqn (3.5) is valid if d$\boldsymbol{l}$ is simply a line element of a purely mathematical closed path of integration. The only requirement is that the path should be fixed. If we apply (3.5) to an infinitesimal loop of infinitesimal area, the definition of the curl of a vector, i.e. $\nabla \wedge \boldsymbol{E}$ allows us to write (3.5) as a differential equation

$$\nabla \wedge \boldsymbol{E} = -\dot{\boldsymbol{B}}. \tag{3.6}$$

This is our third Maxwell equation. It expresses an intrinsic property of the electromagnetic field—changing magnetic fields generate electric fields. In the next chapter we shall see that changing electric fields also generate magnetic fields. It is no coincidence that we would have expected (3.5) from general ideas about relativity. Eqn (3.6) is one of the field equations that leads to the conclusion that the velocity of light in vacuum is independent of the motion of the source or the observer, and this in turn is the foundation of special relativity.

Eqns (3.5) or (3.6) describe the fundamental effects utilized in transformers but their consequences are most clearly exhibited in the betatron particle accelerator. In this device electrons perform circular orbits in a plane normal to a magnetic field. As the field is increased the flux within an orbit increases and the electric field acting round the orbit, which may be derived from (3.5), accelerates the electrons and increases their kinetic energy.

The measurement of *B*

The flip-coil fluxmeter measures $\boldsymbol{B}$ in terms of the Lorentz force on the charge carriers in the moving flip-coil, and does not involve Faraday's law. If, however, a coil is placed between the poles of an electromagnet which is initially un-energized, so that $\boldsymbol{B} = 0$, and the e.m.f. produced by the coil is integrated electronically as $\boldsymbol{B}$ is increased, we have

$$\int_0^t V\,\mathrm{d}t = -BAN. \tag{3.7}$$

The integrating fluxmeter therefore utilises Faraday's law.

When charges flow in a conductor in a direction normal to a field $\boldsymbol{B}$ the Lorentz force leads to charge separation, and therefore to an e.m.f. transverse to the directions of $\boldsymbol{B}$ and the current flow. this is known as the Hall effect and probes using the Hall effect in semiconductors are now used for most rough work.

On p. 72 we discuss the phenomenon of magnetic resonance. The Lorentz force on the charge of spinning protons causes the axis of spin to precess at an angular rate $\omega = \gamma B$ where γ is the ratio of magnetic moment to spin angular momentum and B is the field acting on the protons. Since γ is known with great accuracy and ω can be measured with even greater accuracy, this technique, with an inherent accuracy of a few parts in a million, is now used in all precise work.

If a small coil of area A and N turns is placed in an alternating magnetic field, whose component normal to the coil is $B_n \cos \omega t$, the induced e.m.f. given by Faraday's law is $\omega A N B_n \sin \omega t$. This can be used to measure, for example, the stray field due to a.c. machinery, or the radio-frequency field due to a distant transmitter. If, of course, a coil is wound around the yoke of a transformer the e.m.f. developed responds to the rate of change of magnetic flux in the yoke.

We see that either the Lorentz force or the Faraday induced e.m.f. are the basis of all common methods of measuring $\boldsymbol{B}$.

Inductance

When the current I in a coil changes, the changing flux results in an induced e.m.f. This is conventionally expressed as a back e.m.f. V, just as the voltage IR across a resistance is a back e.m.f. The coefficient L in the relation $V = L\dot{I}$ is known as the self-inductance of the coil and expressed in henrys (H). In principle we can calculate the inductance of any coil from its geometry by using the field equations. In practice this is very difficult except for simple coils. Thus, if a long solenoid of radius r and length l has n turns per unit length, the field due to a current I is $\mu_0 n I$ and the flux linking the nl turns of the solenoid is $nl\pi r^2 \mu_0 n I$. As I changes, the induced e.m.f. is

$$nl\pi r^2 \mu_0 n \dot{I} = \mu_0 \pi r^2 n^2 l \dot{I}$$

and so the inductance is $L = \mu_0 \pi r^2 n^2 l$. We could in fact verify that L was positive by careful attention to sign conventions but it is easier to remember Lenz' law—the e.m.f. acts in a direction such as to reduce the change in the linked flux and therefore also I. We can give an alternative demonstration that L must be positive. In Fig. 3.2 a current I_0 has been set up at $t = 0$ and, since there is no net e.m.f. acting in the circuit, we must have $L\dot{I} + RI = 0$. The solution of this is $I = I_0 \exp(-Rt/L)$. If L is positive the current decays to zero. The total energy dissipated as heat in R is finite. If L were negative the heat would be infinite and the system would be a free source of energy.

The work done to set up the current I_0 in L can be calculated from the back e.m.f. $L\dot{I}$. It is the integral of

$$IL\dot{I} = \frac{\mathrm{d}}{\mathrm{d}t}(\tfrac{1}{2}LI^2)$$

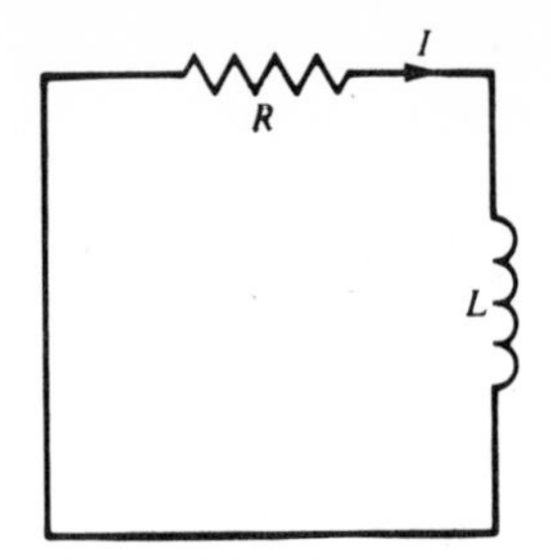

FIG. 3.2. A circuit with resistance and inductance.

and so is $\frac{1}{2}LI_0^2$. This is also the heat dissipated in R as the current decays for this is

$$Q = \int_0^\infty RI^2\,\mathrm{d}t = \int_0^\infty RI_0^2 \exp\left(-\frac{2Rt}{L}\right)\mathrm{d}t = \tfrac{1}{2}LI_0^2.$$

It appears that $\frac{1}{2}LI_0^2$ is the energy initially stored in L. For a long solenoid

$$\tfrac{1}{2}LI^2 = \frac{1}{2\mu_0}\pi r^2 l(\mu_0 nI)^2 = \pi r^2 l \frac{1}{2\mu_0}B^2.$$

Since $\pi r^2 l$ is the volume of the solenoid $B^2/2\mu_0$ can be regarded as the energy density of the magnetic field B. We return to this question in Chapter 7.

A changing current I_1 in one coil induces an e.m.f. V_2 in a nearby coil and vice versa. The coefficients M_{21} and M_{12} in $V_2 = M_{21}\dot{I}_1$ and $V_1 = M_{12}\dot{I}_2$ are known as mutual inductances. Again the unit is the henry. If we first raise I_1 from zero to its final value the stored energy is $\frac{1}{2}L_1I_1^2$ and, when I_2 is then raised from zero, the generator of I_2 does work $\frac{1}{2}L_2I_2^2$ but the induced e.m.f. $M_{12}\dot{I}_2$ in the first coil also requires work $M_{12}I_1I_2$ to maintain I_1 constant. If the sequence of operations is reversed this term is replaced by $M_{21}I_2I_1$. Since the final state is the same we must have $M_{12} = M_{21} = M$ and the stored energy is

$$U = \tfrac{1}{2}L_1I_1^2 + \tfrac{1}{2}L_2I_2^2 + MI_1I_2. \tag{3.8}$$

Mutual inductance can have either sign, we have only to reverse connections on one coil to reverse the sign of M but, since U must be positive for all possible currents, we have

$$M^2 \leqslant L_1L_2. \tag{3.9}$$

A system in which $M^2 = L_1L_2$ is known as a perfect, or ideal, transformer. All the flux linking L_1 also links L_2. In elementary a.c. theory we always assume that transformers are perfect and also that the reactances of the windings ωL_1 and ωL_2, at the operating frequency $\omega/2\pi$, are large compared with any impedances in external circuits. Since the air-cored transformers used in radio-frequency work have $M^2 \ll L_1L_2$ and the iron-cored transformers used in audio work do not always have a large enough reactance to satisfy the second condition, elementary notions can be misleading.

It is possible to express the mutual inductance between two coils in vacuum as

$$M = \frac{\mu_0}{4\pi}\oint_1\oint_2 \frac{\mathrm{d}\boldsymbol{l}_1 . \mathrm{d}\boldsymbol{l}_2}{r(1, 2)} \tag{3.10}$$

where $\mathrm{d}\boldsymbol{l}_1$ and $\mathrm{d}\boldsymbol{l}_2$ are line elements in the two circuits and $r(1, 2)$ is the distance between the elements. This result is known as Neumann's formula.

Faraday's law and Kirchhoff's law

In d.c. circuits the line integral $\oint \boldsymbol{E}.\mathrm{d}\boldsymbol{l}$ taken round a closed loop of a circuit is zero and this leads to the law that the algebraic sum of the e.m.f.s in a closed loop is zero. In a.c. circuits the integral is equal to $-\dot{\Psi}$, where Ψ is the linked flux. We generally avoid this complication by introducing the notion of a back e.m.f. or induced e.m.f. Thus in the circuit of Fig. 3.2 the electric field $\boldsymbol{E}$ is actually zero within the windings of the inductance and the basic circuit equation is $RI + \dot{\Psi} = 0$. However, in writing down $RI + L\dot{I} = 0$, we tacitly assume that the circuit is equivalent to that shown in Fig. 3.3. This assumption

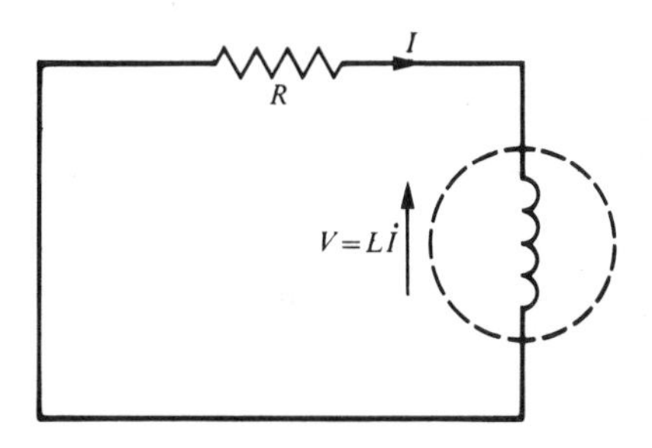

FIG. 3.3. Changing flux in a coil is replaced by a back e.m.f.

causes no confusion in most cases but needs handling with care in high-frequency circuits, especially if the dimensions of the circuit are comparable with a wavelength.

PROBLEMS

3.1. If S is an open surface bounded by a definite fixed perimeter show that $\int_S \boldsymbol{B}.\mathrm{d}\boldsymbol{S}$ is independent of the shape of S.

3.2. If, in Fig. 3.1 the contacts and the wire XYZ rotate with the disc, will there be any e.m.f. across R?

3.3. Calculate the field $\boldsymbol{B}$ at a radius r between the conductors of a coaxial cable with inner and outer conductor radii a and b, when it carries a current I. What is the self-inductance per unit length?

3.4. A current I flows in a long straight wire of radius a. Calculate the energy stored per length in the magnetic field due to the wire. What is the self-inductance per unit length? Why is this absurd result of no physical significance?

3.5. Show that, in the presence of a changing magnetic field, the electric field is no longer a conservative field and cannot be expressed as $-\nabla\phi$.

3.6. A plane coil of N turns and area A has its plane normal to $\boldsymbol{B}$. It is turned over with a uniform angular velocity ω. What is the value of the peak e.m.f. developed in the coil?

3.7. The windings of a bicycle dynamo are fixed to the axle and a permanent magnet to the hub. Is the voltage generated due to Faraday's law?

3.8. A changing current flows in the primary winding of an ideal transformer and the back e.m.f. is V_1. If L_1 and L_2 are the self inductances of the two windings show that the e.m.f. induced in the secondary winding is $V_1(L_2/L_1)^{\frac{1}{2}}$.

3.9. Use eqns (3.8) and (3.10) to calculate the force between two circuits carrying currents I_1 and I_2. Compare your result with eqn (2.8).

3.10. Show that if every linear dimension associated with a coil of wire is doubled, its self-inductance is doubled.

4. Maxwell's Equations. The displacement current

THE equation $\nabla . \boldsymbol{B} = 0$ is a direct mathematical consequence of the expression (2.4b) for the field due to currents in an infinitesimal volume element. This in turn depends on our belief that all magnetic fields are due to currents. Ampère's law $\nabla \wedge \boldsymbol{B} = \mu_0 \boldsymbol{J}$ is on a different footing. We did not derive this law by a direct mathematical argument and the demonstration in terms of currents in a long straight wire is incomplete because we did not discuss what happens at the end of the wire. To see the weakness in this argument consider Fig. 4.1

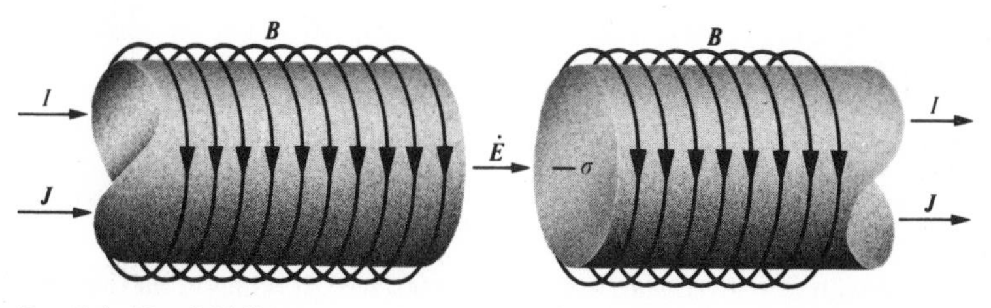

FIG. 4.1. The field lines near a broken conductor carrying an alternating current according to Ampère's law.

in which an alternating current I flows across the capacitance between the ends of two cylindrical conductors. According to Ampère's law the lines of magnetic field terminate at the gap. To Maxwell this seemed entirely unreasonable and he argued as follows. The current lines in the conductors terminate on a changing surface charge σ and if $\boldsymbol{J}$ is the current density then $\dot{\sigma} = \boldsymbol{J}$. The electric field in the gap is $\boldsymbol{E} = \sigma/\epsilon_0$ thus $\epsilon_0 \dot{\boldsymbol{E}} = \boldsymbol{J}$. If then we assume that the changing electric field $\boldsymbol{E}$ produces a magnetic field in exactly the same way as a current density $\boldsymbol{J} = \epsilon_0 \dot{\boldsymbol{E}}$ the magnetic field will continue smoothly across the gap. Maxwell called this current density $\epsilon_0 \dot{\boldsymbol{E}}$ the displacement current density. We shall see in later chapters that it has profound consequences, but we now look at the mathematical reasoning which led Maxwell to this conclusion.

The complete formal mathematical derivation of Ampère's law from eqn (2.4b) depends on the assumption that $\nabla . \boldsymbol{J} = 0$ i.e. that any currents involved flow in closed loops. We can see that this is necessary, for if

$$\boldsymbol{J} = \frac{1}{\mu_0} \nabla \wedge \boldsymbol{B} \quad \text{then} \quad \nabla . \boldsymbol{J} = \frac{1}{\mu_0} \nabla . (\nabla \wedge \boldsymbol{B})$$

is identically zero. Now we know that $\nabla . \boldsymbol{J}$ is not identically zero for if $\nabla . \boldsymbol{J} = 0$, the charge-conservation equation $\nabla . \boldsymbol{J} + \dot{\rho} = 0$ implies that $\dot{\rho} = 0$ and the charge at each point in space is unchangeable. This is clearly false; we can change ρ at any point in space simply by placing a charged body at the point. Thus Ampère's law cannot be a complete statement about the properties of $\nabla \wedge \boldsymbol{B}$. Unfortunately, if in the mathematical derivation from (2.4b) we relax the condition that $\nabla . \boldsymbol{J} = 0$, we can prove nothing more significant than that

$$\nabla . (\nabla \wedge \boldsymbol{B}) = (\nabla . \boldsymbol{J} + \rho)\mu_0.$$

Since we already know that both sides of this equation are zero this is not much use. We note however that in Fig. 4.1 lines of $\boldsymbol{J}$ end on surface charge and lines of $\boldsymbol{E}$ start on the same surface charge, and that if we use $\rho = \epsilon_0 \nabla . \boldsymbol{E}$ to eliminate $\dot{\rho}$ we have $\nabla . \boldsymbol{J} + \dot{\rho} = \nabla . (\boldsymbol{J} + \epsilon_0 \dot{\boldsymbol{E}}) = 0$. Thus we could also prove that

$$\nabla . (\nabla \wedge \boldsymbol{B} - \mu_0 \epsilon_0 \dot{\boldsymbol{E}} - \mu_0 \boldsymbol{J}) = 0.$$

Maxwell's hypothesis is equivalent to the statement that not only is

$$\nabla \wedge \boldsymbol{B} - \mu_0 \epsilon_0 \dot{\boldsymbol{E}} - \mu_0 \boldsymbol{J}$$

a vector with zero divergence but that it is itself zero. It is important to realise that this is a new hypothesis and that its justification is that it leads to results in complete agreement with experiments on a.c. circuits and electromagnetic waves.

We have now completed the formulation of Maxwell's field equations. They are

$$\nabla . \boldsymbol{B} = 0, \tag{4.1}$$

$$\nabla \wedge \boldsymbol{E} + \dot{\boldsymbol{B}} = 0, \tag{4.2}$$

$$\epsilon_0 \nabla . \boldsymbol{E} = \rho, \tag{4.3}$$

$$\nabla \wedge \boldsymbol{B} - \mu_0 \epsilon_0 \dot{\boldsymbol{E}} = \mu_0 \boldsymbol{J}. \tag{4.4}$$

Eqn (4.1) expresses the basic property of the magnetic field $\boldsymbol{B}$ which arises from the way in which it is generated by currents and changing electric fields. The lines of $\boldsymbol{B}$ behave like streamlines in an incompressible fluid and do not end. They can only form closed loops. Eqn (4.2) is Faraday's law of electromagnetic induction. The line integral of $\boldsymbol{E}$ around a closed loop is equal to the negative rate of change of the magnetic flux linking the loop. Eqn (4.3) is essentially Gauss's theorem. It states that lines of $\boldsymbol{E}$ begin and end only on electric charge though they may also form closed loops. Eqn (4.4) is Maxwell's generalization of Ampère's law. The line integral of $\boldsymbol{B}/\mu_0$ round any closed path is equal to the sum of the conduction current and the displacement current linking the path. Notice that eqns (4.1) and (4.2) do not involve charge or current. They express intrinsic properties of the fields. Eqns (4.3) and (4.4) also express properties of the fields if ρ and $\boldsymbol{J}$ are zero, but in addition they express the way in which fields are generated by charges and currents.

In a region where ρ and $\boldsymbol{J}$ are zero the field equations are

$$\nabla . \boldsymbol{B} = 0, \qquad \nabla . \boldsymbol{E} = 0, \tag{4.5a, b}$$

$$\nabla \wedge \boldsymbol{E} = -\dot{\boldsymbol{B}}, \qquad \nabla \wedge \boldsymbol{B} = \mu_0 \epsilon_0 \dot{\boldsymbol{E}}, \tag{4.6a, b}$$

and, in Chapter 10, we show that these equations can be combined to yield two equations for the fields $\boldsymbol{E}$ and $\boldsymbol{B}$ which are

$$\nabla^2 \boldsymbol{E} = \mu_0 \epsilon_0 \ddot{\boldsymbol{E}}, \qquad \nabla^2 \boldsymbol{B} = \mu_0 \epsilon_0 \ddot{\boldsymbol{B}}. \tag{4.7a, b}$$

These are three-dimensional wave-equations for the individual components of $\boldsymbol{E}$ and $\boldsymbol{B}$ and fields which satisfy these equations propagate with a velocity

$$c = (\mu_0 \epsilon_0)^{-\frac{1}{2}}. \tag{4.8}$$

In Maxwell's time electrostatic and magnetic units were defined independently, if not very accurately, and the discovery that the velocity predicted by the field equations for the propagation of electromagnetic waves coincided with the velocity of light was a major triumph for the theory. Today we are so confident that the field equations are correct that we define ϵ_0 as $1/\mu_0 c^2$ and base all our units other than the ampere on this relation.

Maxwell's equations are a complete description of the properties of the electromagnetic fields $\boldsymbol{E}$ and $\boldsymbol{B}$, which exert forces on charges and currents according to the Lorentz law

$$\boldsymbol{f} = \rho \boldsymbol{E} + \boldsymbol{J} \wedge \boldsymbol{B}. \tag{4.9}$$

Eqn (4.9) in conjunction with the field eqns (4.1) to (4.4) gives us a theoretical structure in which only one arbitrary constant μ_0 is required to define electromagnetic units in terms of the velocity of light and mechanical units. We may also add that they need only one additional postulate to contain the theory of special relativity.

PROBLEMS

4.1. Prove that $\nabla . (\nabla \wedge \boldsymbol{B}) \equiv 0$.

4.2. In the figure we show a coaxial capacitor carrying an alternating current $I \cos \omega t$. The whole system is encased in a wax cylinder of diameter 2 in.

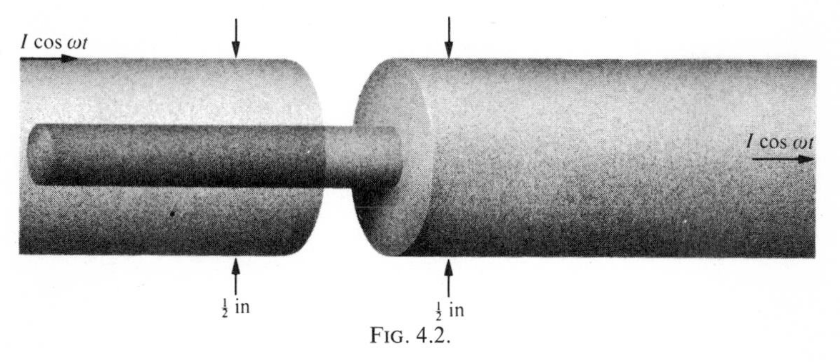

FIG. 4.2.

Is it possible, using a search coil or a magnetic needle, to discover where the capacitor ends and the solid conductor begins by making measurements outside the wax cylinder?

4.3. Two plane parallel plates of area 1 m^2 are placed 1 mm apart and an alternating e.m.f. $V_0 \cos \omega t$ with $V_0 = 1\,000$ volts and $\omega = 2\pi . 10^7 \, s^{-1}$ is applied across the plates. What are the displacement current density and the total displacement current in the space between the plates? Consider a circular path of radius $r = 0{\cdot}1$ m in a plane parallel to the plates in the region between the plates, and use eqn (4.4) (with $\boldsymbol{J} = 0$) to calculate the magnetic field $\boldsymbol{B}$ at a point on this path.

4.4. If $\boldsymbol{B}$ and $\boldsymbol{E}$ are constant in time, show that Maxwell's equations separate into two uncoupled pairs of equations, those of electrostatics and those of magnetostatics.

4.5. Show that Maxwell's equations imply the charge conservation law so that this need no longer be added as a basic equation.

4.6. The capacitor in the tank circuit of a 1 MHz radio transmitter consists of two parallel plates of area 1 m^2 a distance 1 cm apart and the peak voltage across the plates is 10 kV. Find the peak displacement current.

5. Conduction and Ohm's law

IN electrostatics a conductor is a medium containing mobile charge. This charge eventually comes to rest in such a way that the field in the conductor is reduced to zero. In more general terms it is also possible to contemplate a system in which a battery maintains a steady field across a conductor and the charges do not come to rest but instead a steady current flows in the conductor. Experimentally it is then found that, when a potential difference V is established across the ends of a conductor, the steady current I increases with V and in most cases the rate of increase is almost exactly linear. The conductor is then said to obey Ohm's law and the coefficient R in the relation

$$V = IR \tag{5.1}$$

is called the resistance of the conductor. It is also found that the resistance R is proportional to the length of the conductor and varies inversely with its cross-sectional area. This leads to the notion of specific resistivity ρ and conductivity $\sigma = 1/\rho$. (Unfortunately the same letters ρ and σ are commonly used for charge density and surface charge density.) The resistance between the ends of a uniform cylinder of cross-section A and length l is

$$R = \frac{\rho l}{A} = \frac{l}{\sigma A}, \tag{5.2}$$

and the conductance is

$$G = \frac{1}{R} = \frac{\sigma A}{l} = \frac{A}{\rho l}. \tag{5.3}$$

The units of R are ohms and those of conductance reciprocal ohms or mho. (The official SI unit is the siemens but as 'mho' is so much more descriptive we shall use it here.) The units of ρ are ohm metres and the units of σ are mho m^{-1}. The current density $\boldsymbol{J}$ within the conductor satisfies $I = \boldsymbol{J}.\boldsymbol{A}$, and the potential difference is $V = \boldsymbol{E}.\boldsymbol{l}$. We can use this to write Ohm's law in the forms

$$\boldsymbol{J} = \sigma \boldsymbol{E}, \tag{5.4}$$

and

$$\rho \boldsymbol{J} = \boldsymbol{E}. \tag{5.5}$$

We now know that conduction in metals is due to the motion of electrons, and conduction in gases and electrolytes to the motion of electrons and ions. The velocity of a single electron of charge $q = -e$ in a conductor, where there is a field $\boldsymbol{E}$, obeys the equation of motion

$$m\ddot{\boldsymbol{r}} = q\boldsymbol{E} + \boldsymbol{F}(t),$$

where $\boldsymbol{F}$ is the random force due to collisions between the electrons and the atoms of the conductor. The actual motion of the electron is, even setting aside quantum effects, quite complicated and dominated by the random force $\boldsymbol{F}$. The field $\boldsymbol{E}$ has relatively little effect. However, in a metal containing many electrons in the smallest possible volume of practical significance, it is possible to show by detailed analysis that the macroscopic (i.e. laboratory scale) consequences of the collisions can be replaced, in the *average* equation of motion, by a term which looks like friction. If we write this average force as $m\boldsymbol{v}/\tau$ where $\boldsymbol{v}$ is the average drift velocity then τ is related to the mean time between collisions.

The new equation is

$$\dot{\boldsymbol{v}}+\frac{\boldsymbol{v}}{\tau} = \frac{q}{m}\boldsymbol{E}. \tag{5.6}$$

Typically the time τ is of the order of 10^{-12} second or less, in some cases 10^{-16} second, and so, unless the average velocity is changing very rapidly, we can neglect $\dot{\boldsymbol{v}}$ and obtain

$$\boldsymbol{v} = \frac{q\tau}{m}\boldsymbol{E}. \tag{5.7}$$

If there are N electrons in unit volume the current density is $Nq\boldsymbol{v}$ and so

$$\boldsymbol{J} = \frac{Nq^2\tau}{m}\boldsymbol{E} \tag{5.8}$$

and the conductivity in Ohm's law is

$$\sigma = \frac{Nq^2\tau}{m}. \tag{5.9}$$

In metals σ is generally greater than 10^6 mho m^{-1}. In copper it is 6×10^7 mho m^{-1}. The value in semiconductors is much less.

In a conductor the displacement current density $\epsilon_0\dot{\boldsymbol{E}}$, at a frequency $\omega/2\pi$ is of magnitude $\omega\epsilon_0 E$, while the conduction current is σE. The ratio of displacement current to electron current is $\omega\epsilon_0/\sigma \sim 10^{-17}\omega$: thus at all normal frequencies the displacement current in metals can be neglected. If we write $\tau_\mathrm{d} = \epsilon_0/\sigma$, the time τ_d, known as the dielectric relaxation time, has another significance: the equations $\nabla.\boldsymbol{E} = \rho/\epsilon_0$, $\nabla.\boldsymbol{J}+\dot{\rho} = 0$ and $\boldsymbol{J} = \sigma\boldsymbol{E}$ can be combined to give

$$\dot{\rho}+\frac{\rho}{\tau_\mathrm{d}} = \dot{\rho}+\frac{\sigma}{\epsilon_0}\rho = 0 \tag{5.10}$$

and so a transient disturbance of the charge within a metal decays exponentially back to zero in a time $\tau_\mathrm{d} \sim 10^{-17}$ second. From this point of view, since there is effectively no net volume charge, metals obey an electrostatic law at all normal frequencies, and any net charge can only reside on the surface.

If the region between two conducting plates is occupied by a medium with a conductivity σ and the system is regarded as a capacitor the charge on the positive plate q can be calculated as the surface integral, $\int \epsilon_0 E_\mathrm{n}\,\mathrm{d}S$, of the normal component of electric field over its surface. The rate at which charge leaks away is $I = \int \sigma E_\mathrm{n}\,\mathrm{d}S$ and so the charge on the plates obeys the differential equation

$$\dot{q}+I = \dot{q}+\frac{\sigma}{\epsilon_0}q = 0 \qquad (5.11)$$

and the capacitor discharges exponentially with a time constant $\tau_\mathrm{d} = \epsilon_0/\sigma$. Values of τ_d vary from 10^{-17} second for metals to over 10^6 seconds (a week or so) for insulating plastics. There is therefore a range of some 10^{23} to 1 in τ_d between metals and the best insulators. Roughly speaking materials with $\sigma > 10^{-8}$ mho m^{-1} and $\tau_\mathrm{d} < 10^{-3}$ s are regarded as conductors and materials with $\sigma < 10^{-11}$ mho m^{-1} and $\tau_\mathrm{d} > 1$ second are regarded as insulators.

If we have N charged particles per unit volume of charge q and velocity $\boldsymbol{v}$ the rate at which charge crosses an element of area $\boldsymbol{S}$ is $Nq\boldsymbol{v}.\boldsymbol{S}$, so that the associated current density is $\boldsymbol{J} = qN\boldsymbol{v}$. If a field $\boldsymbol{E}$ acts within a volume V the rate at which the kinetic energy of the particles increases is the sum of terms $q\boldsymbol{E}.\boldsymbol{v}$ for each particle in the volume. The rate of increase in unit volume is clearly $Nq\boldsymbol{E}.\boldsymbol{v}$ and we can write this as $\boldsymbol{J}.\boldsymbol{E}$. Thus $\boldsymbol{J}.\boldsymbol{E}$ represents the rate at which the sources of the field $\boldsymbol{E}$, e.g. batteries, are delivering energy to the charge carriers. In a vacuum tube such as a diode this energy becomes the kinetic energy of the electrons, which is eventually dissipated as heat when the electrons strike the anode. In a metal the energy is dissipated as heat in collisions as fast as it is acquired. In both cases, whatever the ultimate fate of the energy as kinetic energy of directed motion or as heat, it is removed from the sources of the field and converted to a new form. If the medium in which the particles move obeys Ohm's law, energy is dissipated as heat at a rate $\sigma E^2 = \rho J^2$ per unit volume but whatever the mechanism the rate is $\boldsymbol{E}.\boldsymbol{J}$. We shall see in Chapter 8 that this is the basis for a discussion of electromagnetic energy.

PROBLEMS

5.1. The resistivity of copper is $1{\cdot}6\times10^{-8}$ ohm metres. Calculate the resistance between two opposing edges of a square sheet of copper 10^{-2} mm thick.

5.2. A typical metal contains about 10^{29} mobile electrons per m^3 and its resistivity is about 5×10^{-7} ohm metres. The charge on the electron is $-1{\cdot}6\times10^{-19}$ coulomb and its mass about 10^{-30} kg. Estimate the time τ which appears in eqn (5.6) and the mean velocity of the electrons when the applied field is 100 V m^{-1}.

5.3. Show that the electric scalar potential in the interior of a conducting medium carrying a current obeys Laplace's equation.

5.4. In the interior of a conducting medium $\boldsymbol{J} = \sigma \boldsymbol{E}$. If σ varies from point to point in the medium is the equation $\sigma \nabla . \boldsymbol{E} + \dot{\rho} = 0$ correct or should one use $\nabla . (\sigma \boldsymbol{E}) + \dot{\rho} = 0$?

5.5. Two copper electrodes are immersed in a liquid whose resistivity is 10^4 ohm m and the resistance measured between the electrodes is found to be 100 ohm. Assume that the relative dielectric constant of the liquid is unity and that $\epsilon_0 \sim 10^{-11}$ F m^{-1}, and calculate the capacitance between the electrodes. If you were to attempt to measure the capacitance using an a.c. bridge what frequency would you choose?

5.6. In general engineering practice a current density of 1 000 A per square inch is regarded as the maximum safe load for copper conductors. How much heat is generated per cubic inch?

6. Dipoles

Introduction

MOST of the systems studied in physics e.g. atoms, crystals or complete circuits have no net charge, but the charge within the system is not uniformly distributed in space and so can give rise to electric effects. The simplest effects are those due to a small relative displacement of the centres of gravity of the positive and negative charges and are equivalent to the effects of a dipole, that is two equal charges of opposite sign a fixed distance apart as shown in Fig. 6.1(a). The product $q\boldsymbol{l}$ is is known as the electric dipole moment $\boldsymbol{p}$. Equally

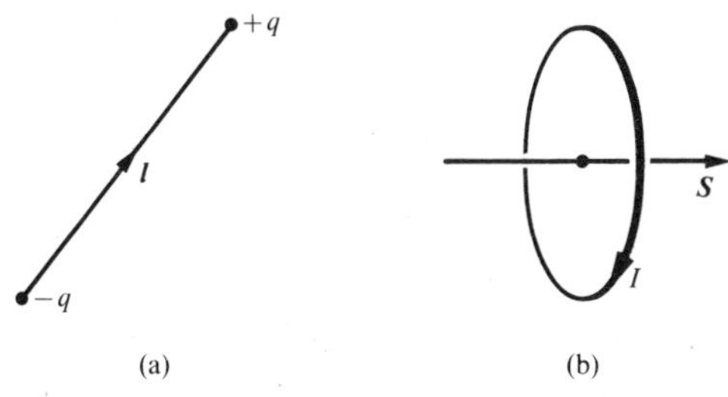

FIG. 6.1. (a) An electric dipole. (b) A magnetic dipole.

the currents in these systems usually form closed loops and do not leave their boundaries. Provided that the systems are also quasi-static, in the sense that charge is not accumulating in any part of the system, the magnetic effects of these currents are equivalent to those of a magnetic dipole i.e. a small loop of wire carrying a steady current. (The notion of a magnetic dipole as a pair of north and south poles at a fixed separation has no place in a discussion of fundamental principles. Not only do magnetic poles not exist but their indiscriminate use is the main reason why so many people find magnetism confusing.) A small plane loop of wire of area S carrying a current I has a magnetic moment IS. If S is treated as a vector the moment is (see Fig. 6.1(b))

$$\boldsymbol{M} = I\boldsymbol{S}. \tag{6.1}$$

We shall prove this result later.

The electric dipole

In a region where the potential is $\phi(\boldsymbol{r})$ due to fixed external charges, the energy of the electric dipole shown in Fig. 6.1(a) is, if the charge $-q$ is at $\boldsymbol{r}$,

$$U = q\{\phi(\boldsymbol{r}+\boldsymbol{l})-\phi(\boldsymbol{r})\}.$$

If ϕ varies slowly, over distances of the order of l, we can express this approximately as

$$U = q\boldsymbol{l}.\nabla_r\phi(\boldsymbol{r})+\tfrac{1}{2}q(\boldsymbol{l}.\nabla_r)^2\phi(\boldsymbol{r})+\text{etc.} \tag{6.2}$$

Only the first item in this series expansion in powers of l will concern us. Since $\boldsymbol{E}(\boldsymbol{r}) = -\nabla_r\phi(\boldsymbol{r})$ this expression can be written in terms of $\boldsymbol{p}$ as

$$U = \boldsymbol{p}.\nabla\phi = -\boldsymbol{p}.\boldsymbol{E}. \tag{6.3}$$

If θ is the angle between $\boldsymbol{p}$ and $\boldsymbol{E}$ this gives

$$U = -pE\cos\theta \tag{6.4}$$

and there is a couple

$$\Gamma = +\frac{\partial U}{\partial\theta} = pE\sin\theta \tag{6.5}$$

acting in a direction which tends to reduce θ. In vector notation this is

$$\boldsymbol{\Gamma} = \boldsymbol{p}\wedge\boldsymbol{E}. \tag{6.6}$$

If $\boldsymbol{E}$ is uniform there is no net force on the dipole but, if $\boldsymbol{E}$ varies slowly with position, the force is

$$\boldsymbol{F} = -\nabla(-\boldsymbol{p}.\boldsymbol{E}).$$

This can be simplified using the vector identity

$$\nabla(\boldsymbol{p}.\boldsymbol{E}) = (\boldsymbol{p}.\nabla)\boldsymbol{E}+(\boldsymbol{E}.\nabla)\boldsymbol{p}+\boldsymbol{p}\wedge(\nabla\wedge\boldsymbol{E})+\boldsymbol{E}\wedge(\nabla\wedge\boldsymbol{p}) \tag{6.7}$$

and, noting that $\boldsymbol{p}$ is a constant and $\nabla\wedge\boldsymbol{E} = 0$, this gives

$$\boldsymbol{F} = (\boldsymbol{p}.\nabla)\boldsymbol{E}. \tag{6.8}$$

We shall show, in a later chapter, that this is also correct even if $\boldsymbol{p}$ is not a constant but varies with $\boldsymbol{E}$.

If we take the origin at the position of the negative charge, the electrostatic potential due to the dipole at a point $\boldsymbol{r}$ is

$$\phi(\boldsymbol{r}) = \frac{q}{4\pi\epsilon_0}\left\{\frac{1}{|\boldsymbol{r}-\boldsymbol{l}|}-\frac{1}{r}\right\},$$

and, since when $r \gg l$ we have

$$\frac{1}{|\boldsymbol{r}-\boldsymbol{l}|} \approx \frac{1}{r}-\boldsymbol{l}.\nabla_r\left(\frac{1}{r}\right), \tag{6.9}$$

this gives the potential at a distance as

$$\phi(\boldsymbol{r}) = -\frac{q}{4\pi\epsilon_0}\boldsymbol{l}.\nabla_r\left(\frac{1}{r}\right) = -\frac{\boldsymbol{p}}{4\pi\epsilon_0}.\nabla_r\left(\frac{1}{r}\right). \tag{6.10}$$

This can also be written in terms of the field $\boldsymbol{e}$ due to unit charge at the origin as

$$\phi(\boldsymbol{r}) = \boldsymbol{p}.\boldsymbol{e}(\boldsymbol{r}). \tag{6.11}$$

The electric field $\boldsymbol{E}(\boldsymbol{r})$ due to the dipole can be obtained using the identity (6.7) and noting that

$$\boldsymbol{\nabla} \wedge \left(\boldsymbol{\nabla}\left(\frac{1}{r}\right)\right) = 0.$$

The result is

$$\boldsymbol{E}(\boldsymbol{r}) = \frac{1}{4\pi\epsilon_0}(\boldsymbol{p}.\boldsymbol{\nabla})\boldsymbol{\nabla}\left(\frac{1}{r}\right) = \frac{3(\boldsymbol{p}.\boldsymbol{r})\boldsymbol{r} - r^2\boldsymbol{p}}{4\pi\epsilon_0 r^5}. \tag{6.12}$$

Notice that, unlike the field of a point charge, this decreases as $1/r^3$. On the axis of the dipole $\boldsymbol{p}.\boldsymbol{r} = pr$ and so the field is $2p/4\pi\epsilon_0 r^3$ and parallel to $\boldsymbol{p}$. In the equatorial plane $\boldsymbol{p}.\boldsymbol{r} = 0$ and the field is $p/4\pi\epsilon_0 r^3$ and opposed to $\boldsymbol{p}$.

If we take spherical polar coordinates r, θ, φ and choose the polar axis to be parallel to $\boldsymbol{p}$ then we have

$$\phi(r, \theta, \varphi) = \frac{p \cos\theta}{4\pi\epsilon_0 r^2}. \tag{6.13}$$

This has the same angular symmetry as the potential

$$\phi = -Er\cos\theta \tag{6.14}$$

which corresponds to a uniform field E parallel to the polar axis and $\boldsymbol{p}$. This relation is useful in many problems.

The magnetic dipole

If, in Fig. 6.1(b), we take the plane of the current loop to be the xy plane so that vector area $\boldsymbol{S}$ is parallel to the z-axis (when the current flows in the sense $x \to y \to -x \to -y \to x$) then in a field $\boldsymbol{B}$ the force acting on an element $\mathrm{d}\boldsymbol{r}$ of the loop is $I\,\mathrm{d}\boldsymbol{r} \wedge \boldsymbol{B}$ and $\mathrm{d}\boldsymbol{r}$ has only components $\mathrm{d}x$ and $\mathrm{d}y$. The net force acting on the whole loop is

$$\boldsymbol{F} = I\oint \mathrm{d}\boldsymbol{r} \wedge \boldsymbol{B}, \tag{6.15}$$

or, if $\boldsymbol{B}$ is uniform,

$$\boldsymbol{F} = I\boldsymbol{B} \wedge \oint \mathrm{d}\boldsymbol{r} = 0,$$

since $\oint \mathrm{d}\boldsymbol{r}$ is zero. The couple acting about the origin is

$$\boldsymbol{\Gamma} = \oint \boldsymbol{r} \wedge (I\,\mathrm{d}\boldsymbol{r} \wedge \boldsymbol{B}). \tag{6.16}$$

The vector identity

$$\boldsymbol{r} \wedge (\mathrm{d}\boldsymbol{r} \wedge \boldsymbol{B}) = (\boldsymbol{r}.\boldsymbol{B})\,\mathrm{d}\boldsymbol{r} - \boldsymbol{B}(\boldsymbol{r}.\mathrm{d}\boldsymbol{r}) \tag{6.17}$$

can be used to simplify (6.16). If $\boldsymbol{B}$ is uniform the last term on the right of (6.17) gives a contribution

$$I\boldsymbol{B}\oint \boldsymbol{r}.\mathrm{d}\boldsymbol{r} = \tfrac{1}{2}I\boldsymbol{B}\oint \mathrm{d}(r^2) = 0, \tag{6.18}$$

since r^2 has the same value at both ends of the path of integration. Remembering that $\boldsymbol{r}$ and $\mathrm{d}\boldsymbol{r}$ lie in the xy plane we see that

$$\boldsymbol{\Gamma} = IB_x\oint x\,\mathrm{d}\boldsymbol{r} + IB_y\int y\,\mathrm{d}\boldsymbol{r}. \tag{6.19}$$

The z-component of $\boldsymbol{\Gamma}$ vanishes because $\mathrm{d}\boldsymbol{r}$ (in the xy plane) has no z-component. In addition $\oint x\,\mathrm{d}x = \frac{1}{2}\oint \mathrm{d}x^2 = 0$ and also $\oint y\,\mathrm{d}y = 0$ while, see Fig. 6.2,

$$\oint x\,\mathrm{d}y = -\oint y\,\mathrm{d}x = S$$

the area of the loop. Thus $\Gamma_x = -ISB_y$ and $\Gamma_y = +ISB_x$. We see that we can write the couple in vector notation as

$$\boldsymbol{\Gamma} = I\boldsymbol{S} \wedge \boldsymbol{B}. \tag{6.20}$$

If we compare this with eqn (6.6) for the couple acting on an electric dipole we see that the loop behaves like a dipole of moment

$$\boldsymbol{m} = I\boldsymbol{S}. \tag{6.21}$$

The magnetic field $\boldsymbol{B}(\boldsymbol{R})$ at a point $\boldsymbol{R}$ due to a small current loop near the origin can be calculated in a number of ways. The method that we shall employ, though not the simplest, has the advantage that it can be immediately generalized to systems more complicated than a plane loop of wire e.g.

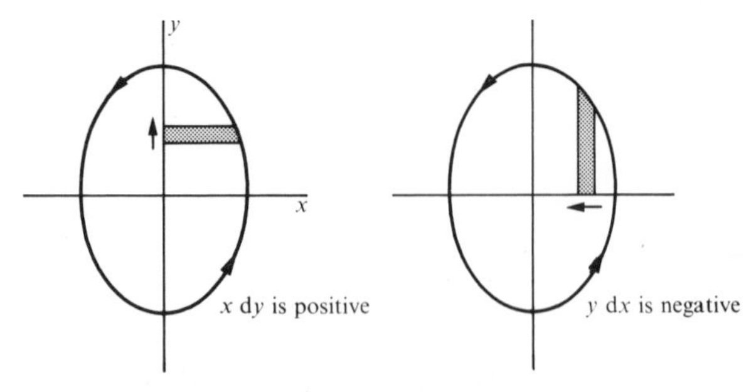

FIG. 6.2. The evaluation of an area.

circulating currents in atoms. We saw in Chapter 2 (see eqn (2.11)) that $\boldsymbol{B}$ could be expressed as $\nabla \wedge \boldsymbol{A}$. The expression for the vector potential $\boldsymbol{A}$, when we are dealing with current in a wire, rather than a more general distribution (see eqn (2.13)) is

$$\boldsymbol{A}(\boldsymbol{R}) = \frac{\mu_0 I}{4\pi} \oint \frac{\mathrm{d}\boldsymbol{r}}{|\boldsymbol{R}-\boldsymbol{r}|}. \tag{6.22}$$

If we write $1/|\boldsymbol{R}-\boldsymbol{r}|$ as $(1/R)-\boldsymbol{r}\,.\,\nabla_R(1/R)$ the integral involving only $1/R$ vanishes since $\oint \mathrm{d}\boldsymbol{r} = 0$. Because the loop lies in the xy plane and $\mathrm{d}\boldsymbol{r}$ has no z-component, $A_z = 0$ and, using the same arguments that lead to eqn (6.20) but with $\boldsymbol{B}$ replaced by $-\nabla_R(1/R)$, we obtain the components of $\boldsymbol{A}(\boldsymbol{R})$ as

$$A_x(\boldsymbol{R}) = +\frac{\mu_0 I}{4\pi} S \frac{\partial}{\partial Y}\left(\frac{1}{R}\right),$$

$$A_y(\boldsymbol{R}) = -\frac{\mu_0 I}{4\pi} S \frac{\partial}{\partial X}\left(\frac{1}{R}\right).$$

With S in the xy plane, it has only the component S_z and so we can reconstruct these relations as a vector relation

$$\boldsymbol{A}(\boldsymbol{R}) = -\frac{\mu_0 I}{4\pi} \boldsymbol{S} \wedge \nabla_R\left(\frac{1}{R}\right), \tag{6.23}$$

or, in terms of $\boldsymbol{m}$,

$$\boldsymbol{A}(\boldsymbol{R}) = -\frac{\mu_0}{4\pi} \boldsymbol{m} \wedge \nabla_R\left(\frac{1}{R}\right). \tag{6.24}$$

The magnetic field $\boldsymbol{B}$ can be obtained using the vector identity

$$\nabla \wedge (\boldsymbol{F} \wedge \boldsymbol{G}) = \boldsymbol{F}(\nabla\,.\,\boldsymbol{G}) - \boldsymbol{G}(\nabla\,.\,\boldsymbol{F}) + (\boldsymbol{G}\,.\,\nabla)\boldsymbol{F} - (\boldsymbol{F}\,.\,\nabla)\boldsymbol{G},$$

with $\boldsymbol{F} = -(\mu_0/4\pi)\boldsymbol{m}$ and $\boldsymbol{G} = \nabla(1/R)$. This is simpler than it might appear, for $\boldsymbol{m}$ is independent of $\boldsymbol{R}$ and $\nabla\,.\,\boldsymbol{G} = \nabla^2(1/R) = 0$, so that

$$\boldsymbol{B} = \nabla \wedge \boldsymbol{A} = -(\boldsymbol{F}\,.\,\nabla)\boldsymbol{G} = \frac{\mu_0}{4\pi}(\boldsymbol{m}\,.\,\nabla)\nabla\left(\frac{1}{R}\right). \tag{6.25}$$

If we compare this with eqn (6.12) for the electric dipole field we see that the two expressions have the same form. Thus the couple acting on the loop and the field produced by the loop both agree with our assumption that it behaves like a dipole of moment $\boldsymbol{m} = I\boldsymbol{S}$.

If $\boldsymbol{B}$ has the same form as $\boldsymbol{E}$, which was derived from a scalar potential, then $\boldsymbol{B}$ can also be derived from a scalar potential i.e.

$$\boldsymbol{B} = -\nabla\phi_m \tag{6.26}$$

and

$$\phi_m(\boldsymbol{R}) = -\frac{\mu_0}{4\pi}\boldsymbol{m}\,.\,\nabla\left(\frac{1}{R}\right). \tag{6.27}$$

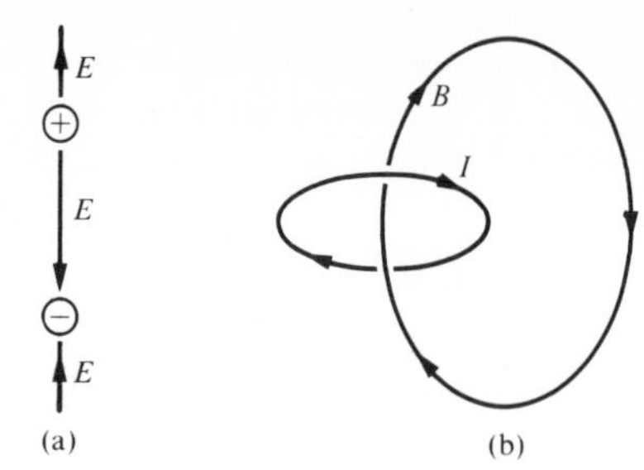

FIG. 6.3. Fields near (a) electric and (b) magnetic dipoles.

This is useful in many calculations and is used in several texts to derive the form of $\boldsymbol{B}$ due to a current loop. It is, however, not quite so straightforward as it might appear. For if $\boldsymbol{B}$ can be written as $-\nabla\phi_m$ the line integral of $\boldsymbol{B}$ round any closed path, even one linking the current loop, is zero. The corresponding result for the electric dipole is correct for, see Fig. 6.3(a), $\boldsymbol{E}$ reverses sign within the dipole, but, see Fig. 6.3(b), it violates Ampère's law in the magnetic case. The expressions for both $\boldsymbol{E}$ and $\boldsymbol{B}$ derived from the scalar potential are both valid for the field at a distance but neither expression is valid near the dipole, and, in converting the expression for $\boldsymbol{B}$ as $\nabla \wedge \boldsymbol{A}$ into $\boldsymbol{B} = -\nabla\phi_m$ we used the relation $\nabla^2(1/R) = 0$, this is only correct for $R \neq 0$. At the dipole, i.e. as $R \to 0$, $\nabla^2(1/R)$ is singular (infinite) and so the expression $\boldsymbol{B} = -\nabla\phi_m$ is completely wrong in this region. The widespread use of eqn (6.27) in discussing the magnetic dipole is, like the use of magnetic poles, a major cause of confsuion. Eqns (6.26) and (6.27), though often useful in calculations of the distant field, must be used with great care. The similarity between the distant fields due to a current loop and a magnetic shell or dipole layer analogous to an electric shell does not imply their complete equivalence. Although Figs. 6.4(a) and 6.4(c) are consistent with the laws of electromagnetism Fig. 6.4(b) is not. It violates the equation $\nabla . \boldsymbol{B} = 0$. The notion that atoms in magnetic materials behave like magnetic shells is both misleading and incorrect.

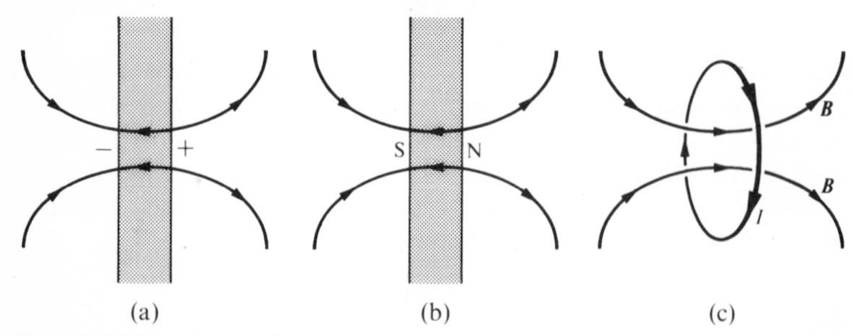

FIG.6.4. (a) The electric shell or dipole layer. (b) The non-existent magnetic shell or dipole layer. (c) The magnetic dipole.

The electric dipole moment of a charge distribution

If a charge distribution $\rho(\boldsymbol{r})$ is confined to a region of linear dimensions l it can be represented by an equivalent charge

$$q = \int_V \rho(\boldsymbol{r})\,\mathrm{d}V(\boldsymbol{r}) \tag{6.28}$$

located at a point $\boldsymbol{R}$ in the charge distribution and an equivalent dipole of moment

$$\boldsymbol{p} = \int_V (\boldsymbol{r}-\boldsymbol{R})\rho(\boldsymbol{r})\,\mathrm{d}V(\boldsymbol{r}) \tag{6.29}$$

also at $\boldsymbol{R}$. Before we show this we note that, if $q = 0$, the value of $\boldsymbol{p}$ is independent of the choice of $\boldsymbol{R}$ even if it is not located within the distribution. If the potential due to external charges is $\phi(\boldsymbol{r})$ and varies slowly over the distribution the potential energy is

$$U = \int_V \rho(\boldsymbol{r})\phi(\boldsymbol{r})\,\mathrm{d}V(\boldsymbol{r}) \approx \int_V \rho(\boldsymbol{r})\{\phi(\boldsymbol{R})+(\boldsymbol{r}-\boldsymbol{R}).\nabla_{\boldsymbol{R}}\phi(\boldsymbol{R})\}\,\mathrm{d}V(\boldsymbol{r}),$$

so that

$$U = \phi(\boldsymbol{R})\int_V \rho(\boldsymbol{r})\,\mathrm{d}V(\boldsymbol{r}) - \boldsymbol{E}(\boldsymbol{R}).\int_V (\boldsymbol{r}-\boldsymbol{R})\rho(\boldsymbol{r})\,\mathrm{d}V(\boldsymbol{r}), \tag{6.30}$$

or

$$U = q\phi(\boldsymbol{R}) - \boldsymbol{p}.\boldsymbol{E}(\boldsymbol{R}). \tag{6.31}$$

Thus in this respect the distribution is described by its charge q and moment $\boldsymbol{p}$.

The potential at $\boldsymbol{R}'$ due to the charge is

$$\phi(\boldsymbol{R}') = \int_V \frac{\rho(\boldsymbol{r})\,\mathrm{d}V(\boldsymbol{r})}{4\pi\epsilon_0\,|\boldsymbol{R}'-\boldsymbol{r}|} = \int_V \frac{\rho(\boldsymbol{r})\,\mathrm{d}V(\boldsymbol{r})}{4\pi\epsilon_0\,|\boldsymbol{R}'-\boldsymbol{R}-(\boldsymbol{r}-\boldsymbol{R})|}$$

and this is approximately

$$\phi(\boldsymbol{R}') = \int_V \frac{\rho(\boldsymbol{r})}{4\pi\epsilon_0}\left\{\frac{1}{|\boldsymbol{R}'-\boldsymbol{R}|} - (\boldsymbol{r}-\boldsymbol{R}).\nabla_{\boldsymbol{R}'}\left(\frac{1}{|\boldsymbol{R}'-\boldsymbol{R}|}\right)\right\}\mathrm{d}V(\boldsymbol{r})$$

or

$$\phi(\boldsymbol{R}') = \frac{q}{4\pi\epsilon_0\,|\boldsymbol{R}'-\boldsymbol{R}|} - \frac{1}{4\pi\epsilon_0}\boldsymbol{p}.\nabla_{\boldsymbol{R}'}\left(\frac{1}{|\boldsymbol{R}'-\boldsymbol{R}|}\right). \tag{6.32}$$

If we compare this with eqn (6.10) replacing $\boldsymbol{r}$ by $\boldsymbol{R}'-\boldsymbol{R}$ we see that this is the potential due to a charge q and a dipole $\boldsymbol{p}$ at $\boldsymbol{R}$.

In atomic physics there is a natural origin to choose for $\boldsymbol{R}$, i.e. the centre of the nucleus, and, since the positive nuclear charge is located very near to this origin, the dipole moment can be calculated from the electronic charge distribution alone

$$\boldsymbol{p} = \int_{\text{atom}} \boldsymbol{r}\rho(\text{electrons})\,\mathrm{d}V. \tag{6.33}$$

In molecules, and the unit cells of crystals, the natural origin is usually the centre of gravity, but in dealing with an ion e.g. OH^- with a net charge, the position of the origin must be specified explicitly.

The magnet moment of a distributed current

If a uniform magnetic field $\boldsymbol{B}(\boldsymbol{r})$ acts in a region where there is a current of density $\boldsymbol{J}(\boldsymbol{r})$, the force on a volume element is $\boldsymbol{J} \wedge \boldsymbol{B}\,\mathrm{d}V$ and this gives a contribution $\boldsymbol{r} \wedge (\boldsymbol{J} \wedge \boldsymbol{B})\,\mathrm{d}V$ to the couple acting about the origin. The total couple is therefore

$$\boldsymbol{\Gamma} = \int_V \boldsymbol{r} \wedge (\boldsymbol{J} \wedge \boldsymbol{B})\,\mathrm{d}V(\boldsymbol{r}). \tag{6.34}$$

The field at $\boldsymbol{R}$ due the current distribution is $\boldsymbol{\nabla}_R \wedge \boldsymbol{A}(\boldsymbol{R})$ where

$$\boldsymbol{A}(\boldsymbol{R}) = \int_V \frac{\mu_0 \boldsymbol{J}(\boldsymbol{r})\,\mathrm{d}V(\boldsymbol{r})}{4\pi\,|\boldsymbol{R}-\boldsymbol{r}|}, \tag{6.35}$$

and it is quite possible, though difficult and tedious, to show that $\boldsymbol{\Gamma}$ can be expressed as $\boldsymbol{m} \wedge \boldsymbol{B}$ and the field by eqn (6.25) if $\boldsymbol{m}$ is taken to be

$$\boldsymbol{m} = \frac{1}{2}\int_V \boldsymbol{r} \wedge \boldsymbol{J}\,\mathrm{d}V. \tag{6.36}$$

The proof requires not only that dimensions of the region should be small compared with $\boldsymbol{R}$ but also that $\boldsymbol{\nabla}.\boldsymbol{J} = 0$. This means both that no current is leaving the region, and that charge is not accumulating in any part of the region i.e. the lines of $\boldsymbol{J}$ must form closed loops. The proof is not particularly illuminating except as an exercise in vector algebra and so we approach the problem from a different point of view.

The discussion in the section starting on p. 35 dealt only with plane loops of wire and we first have to generalize these results to deal with bent loops, for which the simple concept of area has no meaning. In the case of a plane loop in the xy plane the area can be written as either $-\oint y\,\mathrm{d}x$ or $+\oint x\,\mathrm{d}y$ and therefore also as $\frac{1}{2}\oint x\,\mathrm{d}y - \frac{1}{2}\int y\,\mathrm{d}x$. This is clearly the z-component of a vector integral $\frac{1}{2}\oint \boldsymbol{r} \wedge \mathrm{d}\boldsymbol{r}$ and it is not difficult to see that for any loop, plane or bent, the three components of this integral are the areas of the projections of the loop on the three coordinate planes. It is then possible to retrace all the steps in that section and show that the dipole moment of any current loop is

$$\boldsymbol{m} = \frac{1}{2}\oint I\boldsymbol{r} \wedge \mathrm{d}\boldsymbol{r}. \tag{6.37}$$

If the currents described by $\boldsymbol{J}$ have closed flow lines i.e. if $\nabla . \boldsymbol{J} = 0$ we can trace out, in the flow, closed tubes (of varying cross-section) each of which is equivalent to a single current loop because the total current in each tube is constant. Eqn (6.36) then expresses the sum over all these tubes as an integral. A detailed mathematical investigation does no more than confirm this intuitive idea.

Multipole moments and time-dependent systems

The two moments $\boldsymbol{p} = \int \boldsymbol{r}\rho \,\mathrm{d}V$ and $\boldsymbol{m} = \int \frac{1}{2}\boldsymbol{r} \wedge \boldsymbol{J} \,\mathrm{d}V$ are identified as equivalent to dipoles by arguments which ultimately depend on the assumption that we are dealing with static situations. In a static context it is possible to go on to define more complicated moments which specify the distribution of the charges or currents in finer detail. An example is the tensor electric quadrupole moment which has nine components, a typical one being

$$Q_{xy} = \frac{1}{2}\int \rho(\boldsymbol{r})xy \,\mathrm{d}V.$$

The oxygen molecule O_2, for example, has an electric quadrupole moment even though it has no net charge or dipole moment. In time-dependent situations it is no longer possible to identify the expressions for $\boldsymbol{p}$, $\boldsymbol{m}$ and Q in a direct way. Nevertheless, the expressions themselves still exist and can be calculated if ρ is known. It is also possible to show that if $\boldsymbol{p}$, $\boldsymbol{m}$, Q etc. are regarded as defined by their expression in terms of volume integrals, then all electromagnetic effects, whether static or not can still be expressed in terms of these moments and their time derivatives. Because charge is conserved q cannot change in an isolated or complete system and so the effects of q are essentially confined to electrostatics. For example, the dominant term in any description of radiation by atoms can be expressed in terms of $\dot{\boldsymbol{p}}$ and the next most important term involves $\dot{\boldsymbol{m}}$ and $\dot{Q}$. Much the same remarks could be made about either nuclei emitting γ-rays or aerials emitting radio-waves. Because of this, any discussion of the way atoms in matter affect electromagnetic fields reduces to a discussion of either the effects of mobile charges (electrons in a metal) or the effects of bound atomic charge described in terms of atomic electric and magnetic dipole moments.

PROBLEMS

6.1. A metal sphere of radius b is placed in a region where there is a uniform electric field $\boldsymbol{E}$ due to distant charges. Using a polar coordinate system, with the polar axis parallel to $\boldsymbol{E}$, show that the electrostatic potential outside the sphere (the sphere is at zero potential) is given by $\phi = -Er\cos\theta + (A\cos\theta/r^2)$ and find the value of A which makes the surface of the sphere $r = b$ a zero equipotential. The potential $(A\cos\theta)/r^2$ is the potential of a dipole. What is the value of the dipole moment? Calculate the value of the normal component of the field (E_r) at the surface of the sphere. Use the result to calculate the surface charge density on the sphere and then calculate the induced dipole moment of the sphere.

6.2. What is the energy of interaction of two dipoles of moments $\boldsymbol{p}_1$ and $\boldsymbol{p}_2$ separated by a distance $\boldsymbol{r}$?

6.3. Show that $\int_V \nabla^2(1/r)\,\mathrm{d}V$ is zero if V does not include the origin and -4π if it includes the origin. (Hint apply Gauss's theorem to a charge $q = 4\pi\epsilon_0$ at the origin.) Why does this result show that $\nabla^2(1/r)$ is singular at the origin?

6.4. Sketch the lines of force near a short electric dipole.

6.5. The charge density in a cylindrical region with its axis parallel to x is independent of y and z within the cylinder, whose radius is a. It varies with x as $\rho_0 \sin x$ for $-\pi < x < \pi$ and is zero elsewhere. Calculate the dipole moment of the system.

6.6. Fig. 6.5 shows a loop of wire which follows the sides of a cube of side a and carries a current I. What is the magnetic dipole moment of the system? Would it rotate if a field $\boldsymbol{B}$ were applied parallel to an edge of the cube?

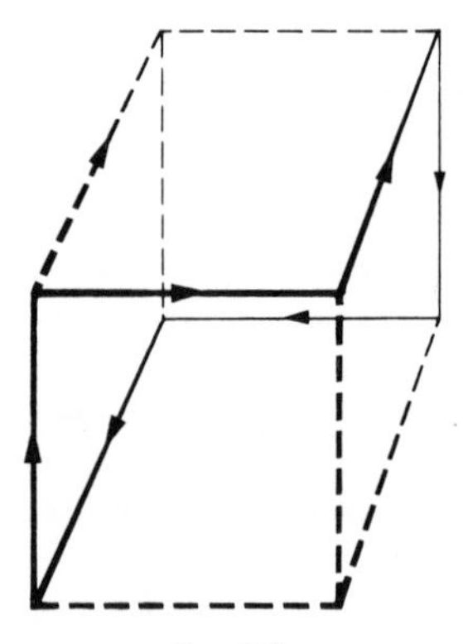

FIG. 6.5.

6.7. Verify, from first principles, the first term on the right of eqn (6.2).

6.8. Verify that eqn (6.6) rather than $\boldsymbol{\Gamma} = \boldsymbol{E} \wedge \boldsymbol{p}$ is correct.

7. Fields in matter

Introduction

IF the space between two isolated charged plates is filled with an insulating medium such as glass or oil the potential difference between the plates decreases. It appears that the medium has reduced the field produced by a given charge. If the plates are parallel and have a surface charge density σ, the field in vacuum is σ/ϵ_0. In the presence of the insulator it is $(\sigma - P)/\epsilon_0$. The quantity P is called the polarization of the medium.

If the space inside a coil of wire carrying a changing current I is filled with a magnetic medium such as iron or a ferrite, the back e.m.f. is increased. It now appears that the medium contributes an extra term to the flux. In a long solenoid with n turns per unit length the magnetic field is $\mu_0 nI$; the medium increases this to $\mu_0(nI + M)$. The quantity M is called the magnetization of the medium.

Neither of these effects is connected with any measurable displacement of charge or additional currents in the medium. They are due to the redistribution within atoms of the charges and currents in the atoms themselves. In this chapter we shall look at how they modify the field equations and in the next chapter we shall discuss their relation to the atomic structure of matter.

Polarization

In vacuum the relation between the electric field and charge is expressed by Gauss's theorem which is equivalent to $\nabla . \epsilon_0 \boldsymbol{E} = \rho$. If this is applied to two parallel plates a distance d apart carrying charges $\pm\sigma$ per unit area, it leads to $E = \sigma/\epsilon_0$. Thus if in the presence of matter we have $\epsilon_0 E = \sigma - P$ we must replace the original relation by

$$\nabla . (\epsilon_0 \boldsymbol{E} + \boldsymbol{P}) = \rho. \tag{7.1}$$

The potential difference between the plates is $\phi = Ed$ and we could investigate the relation between $\boldsymbol{P}$ and $\boldsymbol{E}$ by measuring $\sigma = \epsilon_0 E + P$ as a function of $\phi = Ed$. The result is known as the *constitutive relation* for the medium. For many materials we find that $\boldsymbol{P}$ is proportional to $\boldsymbol{E}$ and this leads to the simple constitutive relation

$$\boldsymbol{P} = \epsilon_0 \chi_e \boldsymbol{E}. \tag{7.2}$$

The dimensionless constant χ_e is called the electric susceptibility of the medium.

If (7.1) is written as

$$\nabla . (\epsilon_0 \boldsymbol{E}) = \rho - \nabla . \boldsymbol{P} = \rho + \rho_b \tag{7.3}$$

it appears that $\rho_b = -\nabla . \boldsymbol{P}$ generates an electric field $\boldsymbol{E}$ in just the same way as a true charge density. We may think of ρ_b as an effective bound charge density. Thus if we have an uncharged, but polarized, body the distant field that it produces will be that of a dipole of moment

$$\boldsymbol{p} = \int \boldsymbol{r}\rho_b \, \mathrm{d}V = -\int \boldsymbol{r}(\nabla . \boldsymbol{P}) \, \mathrm{d}V.$$

The volume integral extends over a region which includes the whole body and its surface lies outside the body, where $\boldsymbol{P} = 0$. It is then easy to show, by taking individual components, that

$$\int \boldsymbol{r}(\nabla . \boldsymbol{P}) \, \mathrm{d}V = -\int \boldsymbol{P} \, \mathrm{d}V,$$

so that the dipole moment is

$$\boldsymbol{p} = \int \boldsymbol{P} \, \mathrm{d}V. \tag{7.4}$$

We see then that we could interpret $\boldsymbol{P}$ as the dipole moment density or dipole moment per unit volume of the body. In the next chapter we show that if we have a body containing N atoms per unit volume each of dipole moment π then $-\nabla . N\pi$ does indeed correspond to an effective charge density in the medium.

We find it useful to introduce a new electric vector, *the displacement*, defined by

$$\boldsymbol{D} = \epsilon_0 \boldsymbol{E} + \boldsymbol{P}. \tag{7.5}$$

This satisfies the equation

$$\nabla . \boldsymbol{D} = \rho, \tag{7.6}$$

where ρ is the mobile or measurable charge density. Whereas lines of $\boldsymbol{E}$ can begin or end on, either mobile charge, or the effective polarization charge ρ_b associated with discontinuities in $\boldsymbol{P}$, lines of $\boldsymbol{D}$ begin and end only on mobile charge. In a simple dielectric in which $\boldsymbol{P} = \epsilon_0 \chi_e \boldsymbol{E}$ we have

$$\boldsymbol{D} = \epsilon_0(1+\chi_e)\boldsymbol{E} = \epsilon_0 \epsilon \boldsymbol{E} \tag{7.7}$$

and the dimensionless factor $\epsilon = 1+\chi_e$ is known as the (relative) dielectric constant. It is important to realize that $\boldsymbol{D}$ is defined by (7.5) and not by (7.7): there are a number of media e.g. piezoelectrics, pyroelectrics and the non-linear media used in laser physics in which $\boldsymbol{P}$ is not linearly related to $\boldsymbol{E}$ and $\boldsymbol{D}$ cannot be written as $\epsilon\epsilon_0 \boldsymbol{E}$.

Magnetization

The back e.m.f. in a coil is related to the changing linked flux by Faraday's law which is equivalent to the equation $\nabla \wedge \boldsymbol{E} = -\dot{\boldsymbol{B}}$ and the field due to a current I is obtained from Ampère's law or the equation $\nabla \wedge \boldsymbol{B} = \mu_0 \boldsymbol{J}$. If filling the coil with a magnetic medium increases $\boldsymbol{B}$ for a given current, e.g. in a long solenoid of n turns per unit length, from $\mu_0 nI$ to $\mu_0(nI+M)$ then we must replace Ampère's law by

$$\nabla \wedge (\boldsymbol{B}-\mu_0 \boldsymbol{M}) = \mu_0 \boldsymbol{J}. \tag{7.8}$$

The magnetization $\boldsymbol{M}$ vanishes in vacuum and we can investigate the relation between $\boldsymbol{M}$ and $\boldsymbol{B}$ by measuring the magnetic flux linking a coil as a function of the current producing the flux. We find it convenient to introduce a new magnetic vector $\boldsymbol{H}$ defined by

$$\boldsymbol{H} = \frac{1}{\mu_0}\boldsymbol{B}-\boldsymbol{M} \tag{7.9}$$

and this vector satisfies the equation

$$\nabla \wedge \boldsymbol{H} = \boldsymbol{J}. \tag{7.10}$$

In this book we call $\boldsymbol{H}$ the (magnetic) intensity: unfortunately many books call $\boldsymbol{H}$ the (magnetic) field. In a long solenoid $H = nI$ and the flux is related to $\boldsymbol{B} = \mu_0(\boldsymbol{H}+\boldsymbol{M})$: it is therefore easier to relate $\boldsymbol{M}$ to $\boldsymbol{H}$ than to $\boldsymbol{B}$ and for this reason the magnetic constitutive relation of a medium generally gives $\boldsymbol{M}$ in terms of $\boldsymbol{H}$. For some media $\boldsymbol{M}$ is more or less proportional to $\boldsymbol{H}$ and the constitutive relation is

$$\boldsymbol{M} = \chi_{\mathrm{m}} \boldsymbol{H}. \tag{7.11}$$

The field $\boldsymbol{B}$ is then related to $\boldsymbol{H}$ by

$$\boldsymbol{B} = \mu_0(1+\chi_{\mathrm{m}})\boldsymbol{H} = \mu_0 \mu \boldsymbol{H}. \tag{7.12}$$

The dimensionless constants χ_{m} and μ are known as the magnetic susceptibility and the (relative) permeability. Again equations (7.11) and (7.12) do *not* define $\boldsymbol{M}$ and $\boldsymbol{B}$. There are many important magnetic media in which $\boldsymbol{M}$ and $\boldsymbol{B}$ are not proportional to $\boldsymbol{H}$ and in, for example, permanent magnet steel both $\boldsymbol{M}$ and $\boldsymbol{B}$ may be opposed, in direction, to $\boldsymbol{H}$.

If we write (7.8) as

$$\nabla \wedge \boldsymbol{B} = \mu_0(\boldsymbol{J}+\nabla \wedge \boldsymbol{M}) = \mu_0(\boldsymbol{J}+\boldsymbol{J}_{\mathrm{b}}) \tag{7.13}$$

we see that $\boldsymbol{J}_{\mathrm{b}} = \nabla \wedge \boldsymbol{M}$ generates a magnetic field in the same way as a true current density. Thus, if we have a magnetized body free of true currents, its magnetic moment will be

$$\boldsymbol{m} = \tfrac{1}{2}\int \boldsymbol{r} \wedge \boldsymbol{J}_{\mathrm{b}}\, \mathrm{d}V = \tfrac{1}{2}\int \boldsymbol{r} \wedge (\nabla \wedge \boldsymbol{M})\, \mathrm{d}V, \tag{7.14}$$

where the surface of the volume of integration lies outside the body in a region where $\boldsymbol{M} = 0$. This rather alarming integral can be simplified, by considering individual components one at a time, and reduces to

$$\int \tfrac{1}{2}\boldsymbol{r} \wedge (\boldsymbol{\nabla} \wedge \boldsymbol{M})\,\mathrm{d}V = \int \boldsymbol{M}\,\mathrm{d}V. \tag{7.15}$$

Thus the magnetic moment of the body is the volume integral of $\boldsymbol{M}$. It is natural to interpret $\boldsymbol{M}$ as the magnetic moment per unit volume or magnetic dipole moment density. In the next chapter we show that, if a body contains N atoms in unit volume with circulating currents which give each atom a magnetic moment $\boldsymbol{\mu}$, then $\boldsymbol{\nabla} \wedge (N\boldsymbol{\mu})$ represents the effect of these currents in generating magnetic fields.

Maxwell's equations

Two of the vacuum field equations i.e.

$$\boldsymbol{\nabla}.\boldsymbol{B} = 0 \text{ (a)} \quad \text{and} \quad \boldsymbol{\nabla} \wedge \boldsymbol{E} + \dot{\boldsymbol{B}} = 0 \text{ (b)} \tag{7.16}$$

do not involve charges and currents and are unaltered in the presence of matter. The other two equations

$$\boldsymbol{\nabla}.(\epsilon_0 \boldsymbol{E}) = \rho \text{ (a)} \quad \text{and} \quad \boldsymbol{\nabla} \wedge \boldsymbol{B} - \mu_0 \epsilon_0 \dot{\boldsymbol{E}} = \mu_0 \boldsymbol{J} \text{ (b)} \tag{7.17}$$

are modified. To ρ we add the term $-\boldsymbol{\nabla}.\boldsymbol{P}$ and to $\boldsymbol{J}$ the term $\boldsymbol{\nabla} \wedge \boldsymbol{M}$ to allow for the effects of bound atomic charges and currents in matter. This is, however, not quite sufficient for, if $\boldsymbol{P}$ is changing with time, the atomic charges are in motion and there is an extra term in their contribution to the current. We can obtain the form of this extra current as follows. Conservation of mobile charge implies that

$$\boldsymbol{\nabla}.\boldsymbol{J} + \dot{\rho} = 0,$$

but now

$$\dot{\rho} = \boldsymbol{\nabla}.(\epsilon_0 \dot{\boldsymbol{E}} + \dot{\boldsymbol{P}})$$

and so

$$\boldsymbol{\nabla}.(\boldsymbol{J} + \epsilon_0 \dot{\boldsymbol{E}} + \dot{\boldsymbol{P}}) = 0.$$

It appears that the displacement-current density $\epsilon_0 \dot{\boldsymbol{E}}$ is to be augmented by $\dot{\boldsymbol{P}}$. (We return to this question in the next chapter.) Thus the pair of equations (7.17) become

$$\boldsymbol{\nabla}.(\epsilon_0 \boldsymbol{E} + \boldsymbol{P}) = \rho \text{ (a)} \quad \text{and} \quad \boldsymbol{\nabla} \wedge \boldsymbol{B} - \mu_0 \epsilon_0 \dot{\boldsymbol{E}} = \mu_0 (\boldsymbol{J} + \dot{\boldsymbol{P}} + \boldsymbol{\nabla} \wedge \boldsymbol{M}) \text{ (b)} \tag{7.18}$$

In terms of the auxiliary vectors

$$\boldsymbol{D} = \epsilon_0 \boldsymbol{E} + \boldsymbol{P} \text{ (a)} \quad \text{and} \quad \boldsymbol{H} = \frac{1}{\mu_0} \boldsymbol{B} - \boldsymbol{M} \text{ (b)} \tag{7.19}$$

the complete set of field equations is

$$\begin{aligned} \nabla . \boldsymbol{B} &= 0 \text{ (a)}, \qquad \nabla \wedge \boldsymbol{E}+\dot{\boldsymbol{B}} = 0 \text{ (b)}, \\ \nabla . \boldsymbol{D} &= \rho \text{ (c)}, \qquad \nabla \wedge \boldsymbol{H}-\dot{\boldsymbol{D}} = \boldsymbol{J} \text{ (d)}, \end{aligned} \tag{7.20}$$

which is the form originally due to Maxwell. We note, however, that these equations cannot be used unless either $\boldsymbol{P}$ and $\boldsymbol{M}$ are known, or can be related to $\boldsymbol{E}$ and $\boldsymbol{H}$ by a pair of constitutive relations such as (7.2) and (7.12). They do, however, completely specify the nature of the four vectors, $\boldsymbol{E}$, $\boldsymbol{B}$, $\boldsymbol{D}$ and $\boldsymbol{H}$ even without these constitutive relations.

The properties of the electric vectors

The two electric vectors $\boldsymbol{E}$ and $\boldsymbol{D}$ satisfy the equations

$$\nabla \wedge \boldsymbol{E} = -\dot{\boldsymbol{B}} \text{ (a)} \quad \text{and} \quad \nabla . \boldsymbol{D} = \rho \text{ (b)} \tag{7.21}$$

and these are valid in any medium and also in a region which includes the boundary between two different materials of different electrical properties. They determine the two most basic properties of $\boldsymbol{E}$ and $\boldsymbol{D}$. Eqn (7.21a), in integral form, states that the line integral of $\boldsymbol{E}$ round any closed path is related to the changing flux linking the path. If this is applied to the path shown in Fig. 7.1, which consists of two finite straight segments τ in two different

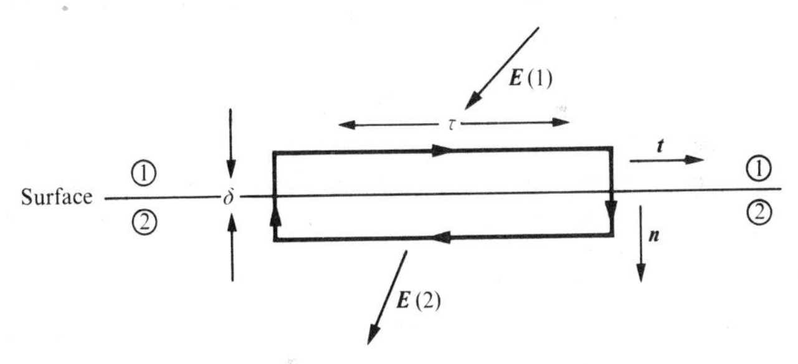

FIG. 7.1. A path of integration lying partly in each of two media.

media and two infinitesimal segments δ normal to the boundary, the flux linking the path tends to zero as $\delta \to 0$ and the integral is $\tau(E_t(1)-E_t(2))$ where E_t is the component of $\boldsymbol{E}$ tangential to the surface. Since τ is finite we have

$$E_t(1) = E_t(2) \tag{7.22}$$

and the tangential component of $\boldsymbol{E}$ is continuous. In electrostatics we can, because $\nabla \wedge \boldsymbol{E} = 0$, express $\boldsymbol{E}$ as $-\nabla\phi$ and, since the normal component of $\boldsymbol{E}$ is $E_n = -\partial\phi/\partial n$ and must be finite, it follows that ϕ is continuous. This is often easier to use than (7.22) in solving for the fields near a boundary.

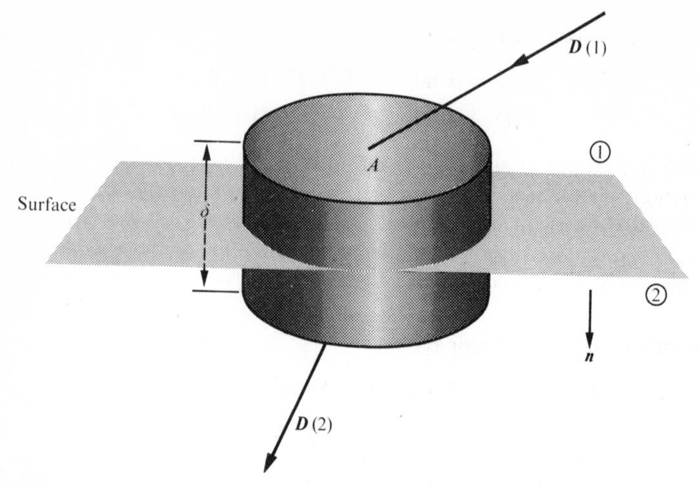

FIG. 7.2. A volume element straddling a surface.

The integral form of (7.21b) relates the surface integral of the normal component of $\boldsymbol{D}$ to the charge within the surface. If this is applied to the flat volume element shown in Fig. 7.2 then, as $\delta \to 0$, the surface integral reduces to $A(D_{\mathrm{n}}(2)-D_{\mathrm{n}}(1))$. If both media are insulators there can be no net mobile charge and so D_{n} is continuous and

$$D_{\mathrm{n}}(1) = D_{\mathrm{n}}(2). \tag{7.23}$$

If one medium is a metal there can be a net surface charge and, because the density of mobile charge in metals is so high, enough charge can accumulate in a layer whose thickness δ is negligible on a practical scale, to invalidate (7.23). We are then left with the relation

$$D_{\mathrm{n}}(2)-D_{\mathrm{n}}(1) = \frac{q}{A} = \sigma, \tag{7.24}$$

which determines the mobile charge σ per unit area of the surface. If medium (1) is a metal, and therefore $\boldsymbol{D}(1) = 0$, we have

$$\sigma = D_{\mathrm{n}}(2). \tag{7.25}$$

At the surface between two insulators $D_{\mathrm{n}}(1) = D_{\mathrm{n}}(2)$ and so

$$\epsilon_0(E_{\mathrm{n}}(1)-E_{\mathrm{n}}(2)) = -(P_{\mathrm{n}}(1)-P_{\mathrm{n}}(2)). \tag{7.26}$$

At a surface where the normal component of $\boldsymbol{P}$ is discontinuous there is an equal and opposite discontinuity in $\epsilon_0 E_{\mathrm{n}}$. Similarly

$$D_{\mathrm{t}}(1)-D_{\mathrm{t}}(2) = P_{\mathrm{t}}(1)-P_{\mathrm{t}}(2) \tag{7.27}$$

and discontinuities in the tangential component of $\boldsymbol{P}$ lead to discontinuities in D_t. These properties reflect the consequences of the equations

$$\epsilon_0 \nabla . \boldsymbol{E} = -\nabla . \boldsymbol{P} \text{ (a)} \quad \text{and} \quad \nabla \wedge \boldsymbol{D} = \nabla \wedge \boldsymbol{P} \text{ (b)}. \tag{7.28}$$

Properties of the magnetic vectors

The field $\boldsymbol{B}$ satisfies the equation $\nabla . \boldsymbol{B} = 0$ and we need not repeat the argument based on Fig. 7.2 to show that at the boundary between two media

$$B_n(1) = B_n(2). \tag{7.29}$$

The intensity $\boldsymbol{H}$ satisfies

$$\nabla \wedge \boldsymbol{H} = \boldsymbol{J} + \dot{\boldsymbol{D}} \tag{7.30}$$

and, if $\boldsymbol{J}$ is everywhere finite this leads to continuity of H_t. We need only replace $\boldsymbol{E}$ by $\boldsymbol{H}$ in Fig. 7.1. However, if one medium is a metal we can easily have a considerable current flowing near the surface in a thin layer whose thickness is negligible on a practical scale. Thus as the thickness δ in Fig. 7.1 is made negligibly small we have

$$\tau(H_t(1) - H_t(2)) = I$$

where I is the surface current linking the path of integration. If $\boldsymbol{K}$ is the surface current per unit length we can express this in vector notation as

$$\boldsymbol{K} = (\boldsymbol{H}(1) - \boldsymbol{H}(2)) \wedge \boldsymbol{n} \tag{7.31}$$

where $\boldsymbol{n}$ is the unit normal from medium (1) to medium (2). If both media are insulators, $\boldsymbol{K}$ of course must be zero and

$$H_t(1) = H_t(2), \tag{7.32}$$

and if one medium is a normal metal $\boldsymbol{K}$ will dissipate energy (ohmic heating) and decay to zero. Thus, at least for static fields, (7.32) is valid unless one medium is a superconductor, in which case we must use (7.31) and H_t may be discontinuous. For oscillatory fields the situation is rather different and we shall discuss this case in Chapter 10.

Because $\nabla . \boldsymbol{B} = \mu_0 \nabla . (\boldsymbol{H} + \boldsymbol{M}) = 0$, we have

$$\nabla . \boldsymbol{H} = -\nabla . \boldsymbol{M}$$

and lines of $\boldsymbol{H}$ may end in discontinuities in $\boldsymbol{M}$. If M_n is discontinuous at a surface we have a discontinuity in H_n given by

$$H_n(1) - H_n(2) = -(M_n(1) - M_n(2)). \tag{7.33}$$

In the absence of surface currents the tangential component of $\boldsymbol{B}$ has discontinuities related to those of M_t i.e.

$$B_t(1)-B_t(2) = \mu_0(M_t(1)-M_t(2)). \tag{7.34}$$

The analogy between magnetic and dielectric media

In static problems, in a medium free of mobile currents, we have $\nabla \wedge \boldsymbol{E} = 0$ and $\nabla \wedge \boldsymbol{H} = 0$. Both $\boldsymbol{E}$ and $\boldsymbol{H}$ can be expressed as gradients of scalar potentials: $\boldsymbol{E} = -\nabla\phi$ and $\boldsymbol{H} = -\nabla\Psi$. If the region is free of mobile charge, we also have both $\nabla.\boldsymbol{D} = 0$ and $\nabla.\boldsymbol{B} = 0$. Thus problems involving dielectric or magnetic media are *formally* the same if we treat $\boldsymbol{E}$ as the analogue of $\boldsymbol{H}$, $\boldsymbol{D}$ as the analogue of $\boldsymbol{B}$ and $\boldsymbol{P}$ as the analogue of $\boldsymbol{M}$. The only difference is the way the constants ϵ_0 and μ_0 appear in the equations. In electrostatics the quantity $\rho_b = -\nabla.\boldsymbol{P}$ behaves in many ways like a charge density and therefore we might complete the analogy by treating $\rho_m = -\nabla.\boldsymbol{M}$ as a magnetic 'pole' density. This is, of course, the basis of many elementary discussions of magnetism and, although it is a useful conceptual aid in some types of calculation, it should be treated with some reserve. Electric charges do exist in the real world but magnetic poles do not. In the next two sections we consider a group of problems involving electric fields whose magnetic equivalents can be written down directly using this analogy.

Fields near dielectric and magnetic bodies

The polarization $\boldsymbol{P}$, of a dielectric body produces fields within and without the body which add to any externally applied field. The general problem of calculating these fields is exceedingly difficult but there are some general results which we now quote without proof. The field $\boldsymbol{E}_1$ in the body due to $\boldsymbol{P}_1$ is uniform if, and only if, $\boldsymbol{P}_1$ itself is uniform and the surface of the body is an ellipsoid described by an equation of the form

$$\left(\frac{x}{a}\right)^2+\left(\frac{y}{b}\right)^2+\left(\frac{z}{c}\right)^2 = 1.$$

Further, the field due to $\boldsymbol{P}_1$ only lies along the same direction as $\boldsymbol{P}_1$ if $\boldsymbol{P}_1$ itself is parallel to one of the principal axes x, y, or z of the body. In this case we can express, for example, the field $\boldsymbol{E}_1$ due to a polarization $\boldsymbol{P}_1$ parallel to x as

$$\boldsymbol{E}_1 = -\frac{1}{\epsilon_0}\gamma_x\boldsymbol{P}_1. \tag{7.35}$$

The magnetic equivalent of this result is

$$\boldsymbol{H}_1 = -\gamma_x\boldsymbol{M}_1. \tag{7.36}$$

The parameters γ_x, γ_y and γ_z are known as the shape factors or demagnetizing factors of the ellipsoid. They depend in a complicated way on the ratios of the principal axes of the ellipse but they satisfy the simple relation

$$\gamma_x + \gamma_y + \gamma_z = 1. \tag{7.37}$$

If the ellipsoid is a sphere, i.e. has three equal axes, symmetry tells us that $\gamma_x = \gamma_y = \gamma_z = \frac{1}{3}$ and the shape factor is $\frac{1}{3}$ for any direction. If the ellipsoid has rotational symmetry about, say, the z-axis so that $a = b$ we have $\gamma_x = \gamma_y$ and, if we call these factors for fields normal to the symmetry axis $\gamma_\perp$, and the factor for fields along the axis $\gamma_\parallel$, we have

$$2\gamma_\perp + \gamma_\parallel = 1. \tag{7.38}$$

Thus if we can calculate either $\gamma_\perp$ or $\gamma_\parallel$ we can obtain the other factor.

A long thin circular cylinder or needle of radius a and length $2L$ approximates an ellipsoid. If it is uniformly polarized parallel to its axis the electric field within the ellipsoid arises from the effective charges $q_b = \pm\pi a^2 P_1$ associated with the discontinuity in the normal component of $\boldsymbol{P}_1$ at the ends (see problem 2). Near the centre of the cylinder these produce a field $2\pi a^2 P_1/4\pi\epsilon_0 L^2$ which tends to zero as $a/L \to 0$. Thus for a long thin cylinder $\gamma_\parallel = 0$ and $\gamma_\perp = \frac{1}{2}$.

A flat thin circular disc approximates an ellipsoid and, if it has a uniform polarization normal to the faces, there will be an effective surface charge of density $\sigma_b = \pm P_1$ on these faces. The field inside the disc is therefore $-P_1/\epsilon_0$ and, for a thin flat disc, $\gamma_\parallel = 1$ and $\gamma_\perp = 0$.

If we have an ellipsoidal cavity in a medium in which there is a uniform applied field $\boldsymbol{E}_0$ and a uniform polarization $\boldsymbol{P}_0$, the field in the cavity is the sum of $\boldsymbol{E}_0$ and the effect of the missing polarization $-\boldsymbol{P}_0$ in the cavity. It is therefore

$$\boldsymbol{E}_{\text{int}} = \boldsymbol{E}_0 + \frac{1}{\epsilon_0}\gamma \boldsymbol{P}_0. \tag{7.39}$$

The magnetic result is

$$\boldsymbol{B}_{\text{int}} = \mu_0(\boldsymbol{H}_0 + \gamma \boldsymbol{M}_0). \tag{7.40}$$

For spherical cavities these give

$$\boldsymbol{E}_{\text{int}} = \boldsymbol{E}_0 + \frac{1}{3\epsilon_0}\boldsymbol{P}_0 = \frac{1}{3}\left(2\boldsymbol{E}_0 + \frac{1}{\epsilon_0}\boldsymbol{D}_0\right) \tag{7.41}$$

and

$$\boldsymbol{B}_{\text{int}} = \mu_0(\boldsymbol{H}_0 + \tfrac{1}{3}\boldsymbol{M}_0) = \tfrac{1}{3}(2\mu_0 \boldsymbol{H}_0 + \boldsymbol{B}_0). \tag{7.42}$$

If a body of dielectric constant ϵ_1 and of ellipsoidal shape is placed in a uniform applied field $\boldsymbol{E}$ parallel to a principal axis of the body, the field $\boldsymbol{E}_{\text{int}}$ and polarization $\boldsymbol{P}_1 = \epsilon_0(\epsilon_1 - 1)\boldsymbol{E}_{\text{int}}$ will be uniform in the body and so

$$\boldsymbol{E}_{\text{int}} = \boldsymbol{E}_0 - \frac{\gamma}{\epsilon_0}\boldsymbol{P}_1 = \boldsymbol{E}_0 - \gamma(\epsilon_1 - 1)\boldsymbol{E}_{\text{int}}.$$

The field in the body is therefore

$$\boldsymbol{E}_{\text{int}} = \frac{\boldsymbol{E}_0}{1+\gamma(\epsilon_1-1)}. \tag{7.43}$$

For a sphere this gives

$$\boldsymbol{E}_{\text{int}} = \frac{3}{2+\epsilon_1}\boldsymbol{E}_0. \tag{7.44}$$

We cannot apply these results directly to calculate the field in a cavity in a medium in which $\boldsymbol{P} = (\epsilon_2-1)\epsilon_0\boldsymbol{E}$ for the presence of the cavity makes $\boldsymbol{E}$ and therefore $\boldsymbol{P}$ no longer uniform in the medium. We can, however, proceed as follows. If, in the outer medium we regard $\epsilon_2\epsilon_0$ as replacing ϵ_0, the dielectric constant of the void relative to $\epsilon_2\epsilon_0$ is $1/\epsilon_2$. Thus formula (7.43) can be used if ϵ_1 is replaced by $1/\epsilon_2$ and so

$$\boldsymbol{E}_{\text{int}} = \frac{\boldsymbol{E}_0}{1+\gamma\left(\dfrac{1}{\epsilon_2}-1\right)} = \frac{\epsilon_2\boldsymbol{E}_0}{\epsilon_2+\gamma(1-\epsilon_2)}. \tag{7.45}$$

For a sphere this gives

$$\boldsymbol{E}_{\text{int}} = \frac{3\epsilon_2\boldsymbol{E}_0}{2\epsilon_2+1}. \tag{7.46}$$

for a long needle-shaped cavity parallel to the field with $\gamma = 0$ we have

$$\boldsymbol{E}_{\text{int}} = \boldsymbol{E}_0,$$

and for a flat disc normal to the field with $\gamma = 1$

$$\boldsymbol{E}_{\text{int}} = \epsilon_2\boldsymbol{E}_0 = \boldsymbol{D}_0/\epsilon_0. \tag{7.47}$$

Finally, for a long cylinder with its axis normal to the field so that $\gamma = \frac{1}{2}$

$$\boldsymbol{E}_{\text{int}} = \frac{2\epsilon_2}{\epsilon_2+1}\boldsymbol{E}_0. \tag{7.48}$$

The magnetic equivalent of this result is

$$\boldsymbol{B}_{\text{int}} = \mu_0\frac{2\mu_2}{\mu_2+1}\boldsymbol{H}_0 = \frac{2\boldsymbol{B}_0}{\mu_2+1}. \tag{7.49}$$

We have so far obtained all these results from eqns (7.35) and (7.37) which we stated without proof. In the case of a sphere we can, however, obtain the results more directly. To show this we consider the case of a sphere of radius a and dielectric constant ϵ_1 embedded in a medium in which the dielectric constant is ϵ_2 and there is an applied field $\boldsymbol{E}_0$. Both within the sphere and outside the sphere $\nabla . \boldsymbol{D} = 0$ and, since $\boldsymbol{D} = \epsilon\epsilon_0\boldsymbol{E}$ and ϵ is not a function of

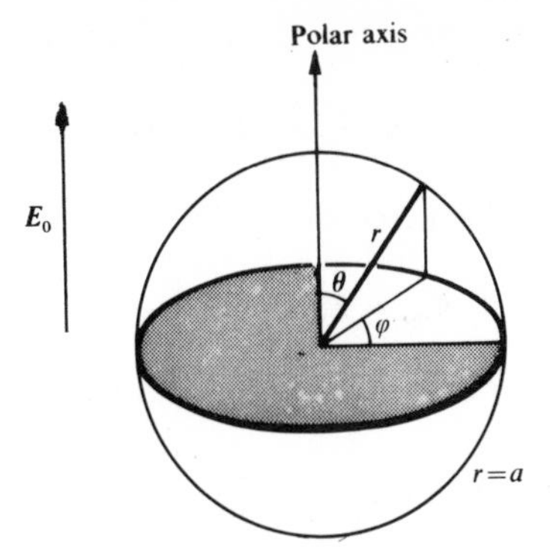

FIG. 7.3. A spherical polar coordinate system.

position, this implies that $\nabla . \boldsymbol{E} = 0$. Thus if within the sphere we write $\boldsymbol{E} = -\nabla\phi_1$ and outside the sphere $\boldsymbol{E} = -\nabla\phi_2$, both the potentials ϕ_1 and ϕ_2 satisfy Laplace's equation $\nabla^2\phi = 0$. We take the spherical coordinate system shown in Fig. 7.3 with the polar axis parallel to $\boldsymbol{E}_0$. The applied field $\boldsymbol{E}_0$ is described by a term $-E_0 r \cos\theta$ in ϕ_2 and, since the fields must match over the surface of the sphere at $r = a$, all other terms in ϕ_1 and ϕ_2 must have the same angular variation. The only solutions of Laplace's equation of this form are $r \cos\theta$ and $\cos\theta/r^2$. Since ϕ_1 must be finite at $r = 0$ it can only be

$$\phi_1 = -E_{\text{int}} r \cos\theta$$

corresponding to a uniform field $\boldsymbol{E}_{\text{int}}$. The exterior potential can contain both terms and is

$$\phi_2 = -E_0 r \cos\theta + \frac{\alpha \cos\theta}{r^2}.$$

Since ϕ_1 must equal ϕ_2 at $r = a$, the constant $\alpha = a^3(E_0 - E_{\text{int}})$ and so

$$\phi_2 = -E_0 r \cos\theta + (E_0 - E_{\text{int}})\frac{a^3 \cos\theta}{r^2}.$$

The normal components of $\boldsymbol{D}$ i.e.

$$D_r = -\epsilon\epsilon_0 \frac{\partial\phi}{\partial r}$$

must be equal at $r = a$ and therefore

$$\epsilon_1 E_{\text{int}} \cos\theta = \epsilon_2(E_0 \cos\theta + 2(E_0 - E_{\text{int}})\cos\theta)$$

so that

$$\boldsymbol{E}_{\text{int}} = \frac{3\epsilon_2}{\epsilon_1 + 2\epsilon_2}\boldsymbol{E}_0. \tag{7.50}$$

We see that this agrees with (7.46) if we set $\epsilon_1 = 1$.

If the exterior medium is vacuum so that $\epsilon_2 = 1$ we have

$$\boldsymbol{E}_{\text{int}} = \frac{3}{2+\epsilon_1}\boldsymbol{E}_0 = \boldsymbol{E}_0 - \frac{\epsilon_1 - 1}{2+\epsilon_1}\boldsymbol{E}_0 = \boldsymbol{E}_0 - \frac{\epsilon_1 - 1}{3}\boldsymbol{E}_{\text{int}}.$$

Since $\boldsymbol{P}_1 = (\epsilon_1 - 1)\epsilon_0 \boldsymbol{E}_{\text{int}}$ this gives

$$\boldsymbol{E}_{\text{int}} = \boldsymbol{E}_0 - \frac{1}{3\epsilon_0}\boldsymbol{P}_1,$$

in agreement with (7.35). Finally if the exterior medium is again vacuum the constant α is

$$a^3(E_0 - E_{\text{int}}) = \frac{a^3}{3}(\epsilon_1 - 1)E_{\text{int}} = \frac{a^3}{3\epsilon_0}P_1.$$

The additional term in the potential outside the sphere is therefore

$$\frac{\alpha \cos\theta}{r^2} = \frac{4\pi}{3}a^3 P_1 \frac{\cos\theta}{4\pi\epsilon_0 r^2} \tag{7.51}$$

which is the potential due to a dipole of moment

$$\boldsymbol{p} = \frac{4\pi}{3}a^3 \boldsymbol{P}_1. \tag{7.52}$$

We can also express this as

$$\boldsymbol{p} = 4\pi a^3 \epsilon_0 \frac{\epsilon_1 - 1}{2+\epsilon_1}\boldsymbol{E}_0$$

and the magnetic equivalent of this is

$$\boldsymbol{m} = 4\pi a^3 \frac{\mu_1 - 1}{2+\mu_1}\boldsymbol{H}_0. \tag{7.53}$$

A superconducting sphere

A superconducting medium not only has zero resistivity but also expels magnetic field. Thus, if we place a superconducting sphere of radius a in a uniform applied field $\boldsymbol{B}_0 = \mu_0 \boldsymbol{H}_0$, circulating currents will be induced in the surface of the sphere which reduce $\boldsymbol{B}$ and $\boldsymbol{H}$ to zero within the sphere. Outside the sphere $\boldsymbol{J} = 0$ and we can express $\boldsymbol{H}$ as $-\nabla\Psi$ and, since $\nabla.\boldsymbol{B} = \mu_0 \nabla.\boldsymbol{H} = 0$, Ψ satisfies Laplace's equation. The potential outside the sphere is therefore

$$\Psi = -H_0 r \cos\theta + \frac{\alpha \cos\theta}{r^2}$$

and the radial component of $\boldsymbol{B}$ is

$$B_r = -\mu_0 \frac{\partial \Psi}{\partial r} = \mu_0\left(H_0 + \frac{2\alpha}{r^3}\right)\cos\theta.$$

Since B_n is continuous this must be zero at the surface of the sphere and so $\alpha = -\frac{1}{2}a^3 H_0$. The tangential component of $\boldsymbol{H}$ just outside the surface of the sphere is

$$H_\theta = -\frac{1}{r}\frac{\partial \Psi}{\partial \theta} = -H_0 \sin\theta + \frac{\alpha}{a^3}\sin\theta = -\tfrac{3}{2}H_0 \sin\theta.$$

Thus, as shown in Fig. 7.4 there must be a circulating current in the surface of the sphere. The contribution between co-latitudes θ and $\theta + \mathrm{d}\theta$ is $H_\theta a\,\mathrm{d}\theta$ and

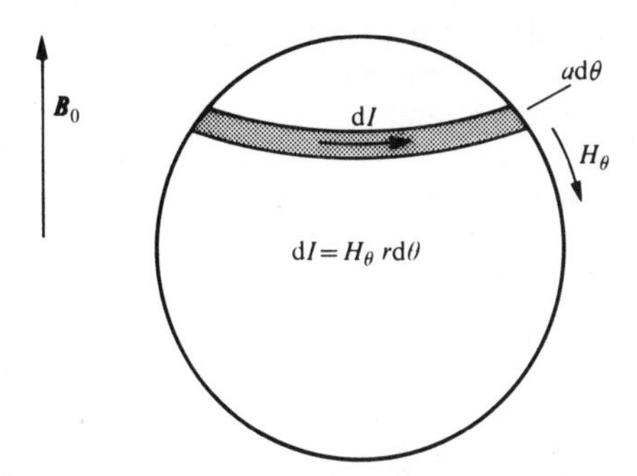

FIG. 7.4. Surface current in a superconducting sphere.

this, circulating in a circle of radius $a \sin\theta$, contributes

$$\mathrm{d}m = H_\theta \pi a^3 \sin^2\theta\,\mathrm{d}\theta = -\tfrac{3}{2}H_0 \pi a^3 \sin^3\theta\,\mathrm{d}\theta$$

to the magnetic moment of the sphere. The total moment is therefore

$$\boldsymbol{m} = -\tfrac{3}{2}\boldsymbol{H}_0 \pi a^3 \int_0^\pi \sin^3\theta\,\mathrm{d}\theta = -2\pi a^3 \boldsymbol{H}_0. \tag{7.54}$$

If we compare this with eqn (7.53) of the last section we see that it is the same as the moment of a sphere of permeability $\mu_1 = 0$ or susceptibility $\chi_\mathrm{m} = -1$. It is, however, not allowable to treat a superconductor as a medium with $\mu = 0$. If we refer back to eqn (7.50) and replace $\boldsymbol{E}$ by $\boldsymbol{H}$ and set $\epsilon_2 = 1$ and $\epsilon_1 = 0$ we get $\boldsymbol{H}_1 = 3\boldsymbol{H}_0/2$. In fact $\boldsymbol{H}_1 = \boldsymbol{B}_1/\mu_0$ is zero in a superconducting sphere.

Forces on charges and currents in matter

The fields $\boldsymbol{E}$ and $\boldsymbol{B}$ are defined so that the forces, in *vacuum*, acting on a charge q and a current element $I\,\mathrm{d}\boldsymbol{l}$ are $q\boldsymbol{E}$ and $I\,\mathrm{d}\boldsymbol{l} \wedge \boldsymbol{B}$. We cannot assume that these expressions hold for charged bodies immersed in dielectric media or wires carrying currents in magnetic media. The questions "What is the force on a charged solid body in a solid dielectric?', 'What is the force on a solid wire in a solid magnetic medium?' are pointless. If the body is at rest the total force is zero and includes a contribution due to mechanical stress in the medium. We cannot even separate out this stress in any unique way, for near a charged body or a wire the fields in the medium are non-uniform, and this itself leads to an additional component of mechanical stress in the medium. The most that we can hope to do is to calculate the additional stress in the medium due to the presence of the body. The case of a body immersed in a fluid medium is, however, quite different. A fluid cannot support a stress and the net body force in the fluid must be zero. This leads to a unique answer for the force on the immersed body and it can be shown quite generally that it is $q\boldsymbol{E}_0$ or $I\,\mathrm{d}\boldsymbol{l} \wedge \boldsymbol{B}_0$ where $\boldsymbol{E}_0$ and $\boldsymbol{B}_0$ are the fields in the medium due to sources other than the body itself. We shall not give the general proof but, instead, give a simple example which shows how it arises. Figure 7.5 shows the

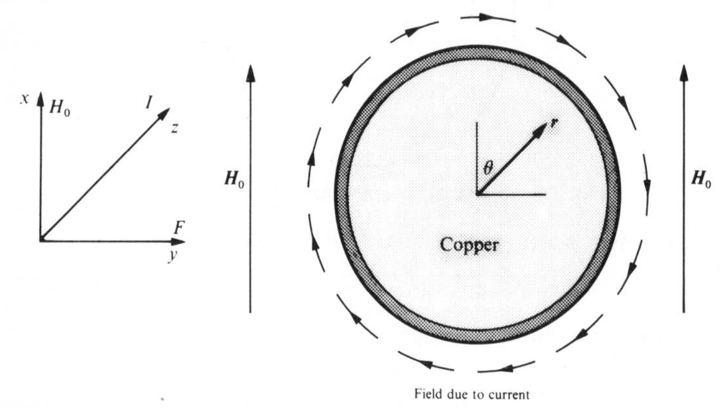

FIG. 7.5. An enamelled wire of circular cross-section in a field.

cross-section of a straight non-magnetic copper wire, with a thin enamel coating, carrying a current I parallel to the z-axis, in a magnetic liquid in which there is a uniform applied field $\boldsymbol{B}_0 = \mu\mu_0\boldsymbol{H}_0$ parallel to the x-axis. The wire forms a cylindrical cavity normal to the field and the field in the cavity is therefore (see 7.49)

$$\boldsymbol{B} = \frac{2}{\mu+1}\boldsymbol{B}_0.$$

The magnetic force acting on unit length of the wire parallel to the y-axis is therefore

$$F'_y = IB = \frac{2}{\mu+1}IB_0.$$

Just outside the surface of the enamelled wire the intensity H due to the current adds to the applied field on the left and subtracts on the right. Thus the field is more intense on the left side and the fluid medium tends to flow into this region. This continues until the resulting compression has set up a pressure gradient which stops further flow.

In the bulk of the fluid this pressure gradient just balances the magnetic force on the fluid but at the surface of the enamelled wire, where μ changes discontinuously to unity, there is an unbalanced force. This acts to attract the fluid towards the centre of the wire and, since it is directed radially inwards is equivalent to a pressure acting on the surface of the wire. The magnitude of the pressure can be shown to be

$$p = \tfrac{1}{2}\mu_0(\mu-1)H^2.$$

If we use cylindrical polar coordinates, as shown in the figure, we have

$$H^2 = H_r^2 + H_\theta^2.$$

The component H_r arises only from H_0 and does not lead to a net force on the wire, but H_θ at the surface of the wire is

$$H_\theta = \frac{I}{2\pi r} - \frac{2\mu}{\mu+1} H_0 \sin\theta$$

and H_θ^2 contains a component

$$H_\theta^{2\prime\prime} = -2\frac{I}{2\pi r}\frac{2\mu}{\mu+1} H_0 \sin\theta$$

which gives

$$p'' = -\mu_0 \frac{\mu-1}{\mu+1}\frac{I}{2\pi r} 2\mu H_0 \sin\theta,$$

and this leads to an additional force on the wire

$$F_y'' = -\int_0^{2\pi} p'' \sin\theta r\, \mathrm{d}\theta = \frac{\mu-1}{\mu+1}\mu\mu_0 H_0 I = \frac{\mu-1}{\mu+1} B_0 I.$$

When this is added to F_y' the total force is

$$F_y = F_y' + F_y'' = \left(\frac{2}{\mu+1} + \frac{\mu-1}{\mu+1}\right) IB_0 = IB_0.$$

We notice the crucial role played by the fluid properties of the medium in leading to this result. Furthermore the mechanical force F_y'' can be seen to be transmitted to the wire through the enamel coating.

Problems of this type are not of much practical interest and we have only given this discussion to show that the question of the force acting on a material body in a material medium is not trivial. The belief that it is, makes it all the more difficult to understand the much more important and complex problem of the forces acting on atoms and electrons in matter to be considered in the next chapter.

PROBLEMS

7.1. If a capacitor is immersed in oil, for which $\boldsymbol{P} = \epsilon_0 \chi_e \boldsymbol{E}$, show that its capacitance is increased by a factor $(1+\chi_e)$.

7.2. A right circular cylinder has a cross section A and length L and a uniform polarization $\boldsymbol{P}$ parallel to its axis. Use eqn (7.3) to show that it generates an electric field as though there were charges $\pm AP$ evenly distributed over its ends. What is the dipole moment of the cylinder?

7.3. A cylinder of cross-sectional area A and length L has a uniform magnetization $\boldsymbol{M}$ parallel to its axis. Use eqn (7.13) to show that it produces the same field $\boldsymbol{B}$ as a solenoid of the same shape with n turns per unit length carrying a current $I = M/n$. If the cylinder is long and thin ($L^2 \gg A$) show that the field on the axis is $\boldsymbol{B} = \mu_0 \boldsymbol{M}$. What is the value of the intensity $\boldsymbol{H}$ on the axis? How does this differ from the value of $\boldsymbol{H}$ in the equivalent solenoid?

7.4. Use eqn (7.22) to show that, at the surface of a perfect conductor $\boldsymbol{E}$ is normal to the surface.

7.5. In vacuum $\boldsymbol{D} = \epsilon_0 \boldsymbol{E}$ and $\epsilon_0 \sim 10^{-11}$ farad m^{-1}. If a field of 10^8 V m^{-1} exists just outside a metal surface, what is the surface charge density on the metal surface? A typical metal contains about 10^{29} mobile electrons, of charge $-1{\cdot}6 \times 10^{-19}$ coulombs, in a cubic metre. How thick, approximately, is the surface charge layer? How does this compare with the thickness of a monatomic layer?

7.6. The electric field in vacuum outside the plane surface of a medium of dielectric constant ϵ makes an angle θ with the normal to the surface. What is the angle between the field and the normal in the medium?

7.7. An anchor-ring of mean circumference L is constructed from a permanent magnet steel which displays pronounced magnetic hysteresis, so that the relation between the magnetization M and the intensity of H forms a closed loop whose equation is

$$\left(\frac{M}{M_0}\right)^2 + \left(\frac{H}{H_0}\right)^2 = 1$$

where M_0 and H_0 are constants. A thin saw cut of thickness t is made across the ring. Draw a diagram showing the directions of H, M and B both in the gap and in the ring. Find the value of B in the gap.

7.8. A thin flat circular disc has a permanent magnetization M normal to its faces. Find the values of $\boldsymbol{B}$ and $\boldsymbol{H}$ within the disc.

7.9. A sphere is constructed of steel in which M and H are related as in problem 7. What are: (1) The intensity $\boldsymbol{H}$ in the sphere? (2) The field $\boldsymbol{B}$ in the sphere? (3) The magnetic moment of the sphere? If a small hole is drilled along the diameter of the sphere parallel to $\boldsymbol{M}$ what is the value of $\boldsymbol{B}$ in the hole?

7.10. An ellipsoid of revolution in which M is related to H as in problem 7 has M parallel to its axis of symmetry. A small hole is drilled along this axis. If the demagnetizing factor is γ what is the value of B in the hole? For what shape of the ellipsoid is B in the hole a maximum?

7.11. A sphere of dielectric constant ϵ is placed in a uniform applied field $\boldsymbol{E}_0$. What is the field in a small spherical cavity at the centre of the sphere?

7.12. An alternating current $I_0 \cos \omega t$ flows in the windings of a long solenoid filled with a medium whose constitutive relation is $M = \chi_m H + \alpha H^3$ where χ_m and α are constants. Show that the back e.m.f. contains a term of angular frequency 3ω.

7.13. Verify the relation between the integrals in eqn (7.15).

7.14. Show that in a system containing only permanent magnets and free of current the total volume integral $\int_V \boldsymbol{H}.\boldsymbol{B}\, \mathrm{d}V$ is zero. (Hint:— express $\boldsymbol{B}$ as $\nabla \wedge \boldsymbol{A}$ and use $\nabla \wedge \boldsymbol{H} = \boldsymbol{J} = 0$ together with the integral form of the relation

$$\nabla.(\boldsymbol{F} \wedge \boldsymbol{G}) = \boldsymbol{G}.(\nabla \wedge \boldsymbol{F}) - \boldsymbol{F}.(\nabla \wedge \boldsymbol{G})).$$

Use your result to show that $\boldsymbol{H}$ is opposed to $\boldsymbol{B}$ within an isolated permanent magnet.

7.15. Show that in a system free of mobile charge $\int_V \boldsymbol{E}.\boldsymbol{D}\, \mathrm{d}V = 0$. Use this result to show that, if a piezoelectric crystal acquires a polarization as a result of strain, the vectors $\boldsymbol{E}$ and $\boldsymbol{D}$ are opposed within the crystal.

7.16. Many crystals, as a result of their molecular structure have an intrinsic permanent polarization $\boldsymbol{P}_0$ of the order of 0·1 coulomb m^{-2}. If a sphere of radius 1 cm is cut from such a crystal, whose dielectric constant is $\epsilon = 4$, what is the electric field at a distance 10 cm from the centre of the sphere in the direction of $\boldsymbol{P}_0$? Why are effects such as this not observed in the laboratory?

8. The atomic origin of dielectric and magnetic properties

Introduction

IN an electric or magnetic field atoms and molecules acquire a net electric or magnetic dipole moment which is usually proportional to the field. If we express these moments as

$$\boldsymbol{p} = \epsilon_0 \alpha_e \boldsymbol{E}^{\text{loc}} \tag{8.1a}$$

$$\boldsymbol{m} = \frac{1}{\mu_0} \alpha_m \boldsymbol{B}^{\text{loc}} \tag{8.1b}$$

where $\boldsymbol{E}^{\text{loc}}$ and $\boldsymbol{B}^{\text{loc}}$ are the local fields at the atomic site, the electric and magnetic polarizabilities α_e and α_m have the dimensions of a volume.

In a medium with N atoms per unit volume the electric and magnetic dipole-moment densities are $N\boldsymbol{p}$ and $N\boldsymbol{m}$. The first problem in a theory of dielectric and magnetic behaviour is to show that these dipole-moment densities, expressed in terms of microscopic, or atomic, variables, are equivalent to the macroscopic vectors $\boldsymbol{P}$ and $\boldsymbol{M}$ used in discussing the bulk properties of matter. The next problem is to relate the fields $\boldsymbol{E}^{\text{loc}}$ and $\boldsymbol{B}^{\text{loc}}$ which act on individual atoms to the bulk, or macroscopic, fields $\boldsymbol{E}$ and $\boldsymbol{B}$ in the medium. The final problem, which is essentially one in atomic physics is to relate the polarizabilities α_e and α_m to the structure of the atoms or molecules.

Macroscopic and microscopic quantities

Even in optics the least linear dimensions which are of practical significance are of the order of a few hundred angströms (1 Å $= 10^{-10}$ m) whereas atomic dimensions are of the order of 1 Å. Thus the variables such as $\boldsymbol{E}$, $\boldsymbol{B}$, $\boldsymbol{P}$, $\boldsymbol{M}$, ρ and $\boldsymbol{J}$ which occur in practical macroscopic problems are, in some sense, averages over regions containing appreciable numbers of atoms. The nature of this average is, however, quite subtle. It is not, as many texts suggest, either a simple volume average or a statistical or quantum mechanical average. To see the nature of the problem we consider a very simple and almost trivially obvious example. Suppose that atoms are arranged on a line with a microscopic spacing a and we wish to define N, the number of atoms per unit length, for use in a macroscopic problem where the smallest distinguishable displacement is Λ which is much greater than a. We proceed by taking an interval of length L which, though greater than a, is less than Λ, and count the number of atoms n in L. If $L = (\nu + \epsilon)a$ where ν is integral and $0 < \epsilon < 1$, the result will be either ν, or $\nu + 1$, depending on the exact location of L

relative to the atoms. This exact location, to within a distance a, has no macroscopic meaning, and so we next average n over all possible positions of L compatible with the macroscopic accuracy inherent in the problem. Clearly the probability that $n = \nu$ is $1-\epsilon$ and that $n = \nu+1$ is ϵ so that the average is $\langle n\rangle = \nu(1-\epsilon)+(\nu+1)\epsilon = \nu+\epsilon$, and $N = \langle n\rangle/L = 1/a$. This gives a definite value for N. We can now define the macroscopic charge and current densities $\langle\rho\rangle$ and $\langle \boldsymbol{J}\rangle$ in terms of the exact microscopic densities ρ and $\boldsymbol{J}$. To define $\langle\rho\rangle$ we choose a volume V, large on a microscopic scale (containing many atoms) yet small on a practical macroscopic scale, and form $1/V \int \rho \, \mathrm{d}V$. We then average this over all locations of V, relative to the atoms, that are macroscopically indistinguishable. To define $\langle \boldsymbol{J}\rangle$ we choose a similar area A_x normal to the x-axis and calculate the current $I = \int J_x \, \mathrm{d}A_x$ crossing this area. The x-component $\langle J_x\rangle$ is then the average of I/A_x when A_x is allowed to take all macroscopically indistinguishable positions relative to the atoms.

The exact microscopic fields $\boldsymbol{e}$ and $\boldsymbol{b}$ are related to the exact charge and current densities ρ and $\boldsymbol{J}$ by the equations $\epsilon_0 \nabla.\boldsymbol{e} = \rho$ and $\nabla \wedge \boldsymbol{b} - \mu_0\epsilon_0\dot{\boldsymbol{e}} = \mu_0\boldsymbol{J}$ and these equations are linear. Thus, if we form the macroscopic averages of ρ and $\boldsymbol{J}$, and so smooth out all fluctuations over microscopic distances leaving only their slow macroscopic variation, the fields $\boldsymbol{E}$ and $\boldsymbol{B}$ that satisfy $\epsilon_0\nabla.\boldsymbol{E} = \langle\rho\rangle$ and $\nabla \wedge \boldsymbol{B} - \mu_0\epsilon_0\dot{\boldsymbol{E}} = \mu_0\langle \boldsymbol{J}\rangle$ will be the required smooth macroscopic fields. As a result we need not consider how these fields are averaged and, when we come to consider the relation between the macroscopic vectors $\boldsymbol{P}$ and $\boldsymbol{M}$ and the atomic polarization and magnetization densities $N\boldsymbol{p}$ and $N\boldsymbol{m}$, we need only show that the effective macroscopic charge and current densities are $\rho_b = -\nabla.(N\boldsymbol{p})$ and $\boldsymbol{J}_b = (\partial/\partial t)(N\boldsymbol{p}) + \nabla \wedge (N\boldsymbol{m})$ to identify $N\boldsymbol{p}$ with $\boldsymbol{P}$ and $N\boldsymbol{m}$ with $\boldsymbol{M}$. This occupies the next two sections.

Electric polarization

To show that $\rho_b = \nabla.(N\boldsymbol{p})$ in three dimensions is beyond the author's artistic capacity and probably beyond the reader's geometrical understanding. We therefore consider the one-dimensional array of dipoles shown in Fig. 8.1. Each dipole consists of a positive and negative charge $\pm q$ with a separation l.

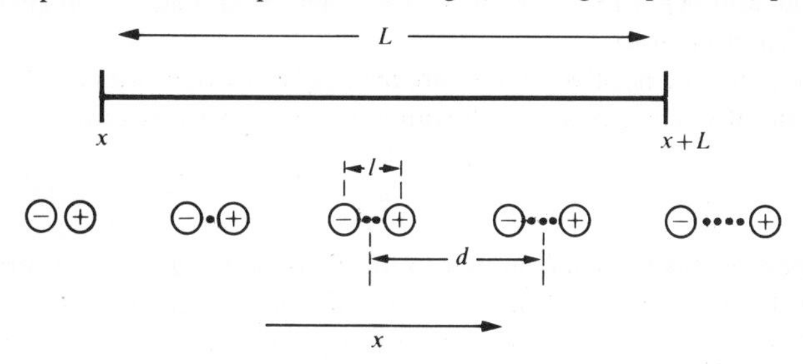

FIG. 8.1. A linear array of atomic dipoles.

This increases towards the right (increasing x) so that the dipole moment also increases. The separation between the dipoles is d, so that the number per unit length is $N = 1/d$, and the local dipole-moment density is $Np = ql/d$. The interval L above the array is of macroscopic length, covering many dipoles, and extends from x to $x+L$. We propose to calculate the average charge in the interval L when its position is random relative to the array of dipoles. The charge will be zero if, either both ends of L fall between dipoles, or both ends fall within dipoles. It will be $+q$ if the left hand end falls within, and the right hand end between, dipoles. The probability of this occurring is

$$\left(\frac{l}{d}\right)_x\left\{1-\left(\frac{l}{d}\right)_{x+L}\right\}.$$

It will be $-q$ if the right hand end falls within a dipole and the left hand end between dipoles, and this has a probability

$$\left(\frac{l}{d}\right)_{x+L}\left\{1-\left(\frac{l}{d}\right)_x\right\}.$$

The expected value of the charge is therefore

$$\langle q\rangle = q\left(\frac{l}{d}\right)_x\left\{1-\left(\frac{l}{d}\right)_{x+L}\right\}-q\left(\frac{l}{d}\right)_{x+L}\left\{1-\left(\frac{l}{d}\right)_x\right\} = \left(\frac{ql}{d}\right)_x-\left(\frac{ql}{d}\right)_{x+L}.$$

We can write this as

$$\langle q\rangle = -\{(Np)_{x+L}-(Np)_x\}.$$

If L, though macroscopic and large on a microscopic scale, is still small on a macroscopic scale and Np varies smoothly with x, we can express this as $-L(\partial/\partial x)(Np)$, so that the expected charge per unit length is

$$\rho_b = \frac{\langle q\rangle}{L} = -\frac{\partial}{\partial x}(Np).$$

The generalization to three dimensions is obvious and so $\rho_b = -\nabla.(N\boldsymbol{p})$ and we can identify $N\boldsymbol{p}$ with $\boldsymbol{P}$, for this vector is defined so that $-\nabla.\boldsymbol{P}$ gives the effective charge density.

If in Fig. 8.1 the positive charges are moving to the right with a velocity v and the negative charges to the left with the same velocity we have

$$\frac{\mathrm{d}p}{\mathrm{d}t} = 2qv$$

and, if now we take a plane normal to the array of charges at x, a positive charge will cross it, going to the right, in $\mathrm{d}t$ if x lies within a distance $v\,\mathrm{d}t$ to the right of a dipole. Similarly a negative charge going to the left will cross it in $\mathrm{d}t$ if x lies within a distance $v\,\mathrm{d}t$ to the left of a dipole. The probability that a randomly placed plane lies in either of these two regions is $v\,\mathrm{d}t/d$ and, since each event corresponds to positive charge moving to the right, the current crossing the plane has an expectation value

$$I = \frac{2qv}{d} = \frac{1}{d}\frac{\mathrm{d}p}{\mathrm{d}t} = \frac{\mathrm{d}}{\mathrm{d}t}(Np).$$

Thus a changing value of Np leads to a current and in three dimensions its density is $(\partial/\partial t)(N\boldsymbol{p})$ which corresponds to the macroscopic term $\dot{\boldsymbol{P}}$.

Magnetization

Fig. 8.2 shows part of the xy plane in a medium in which atomic currents I, circulating in closed loops of area A, give each atom a magnetic moment $m = IA$ directed along the z-axis. The atoms are d apart in rows D apart,

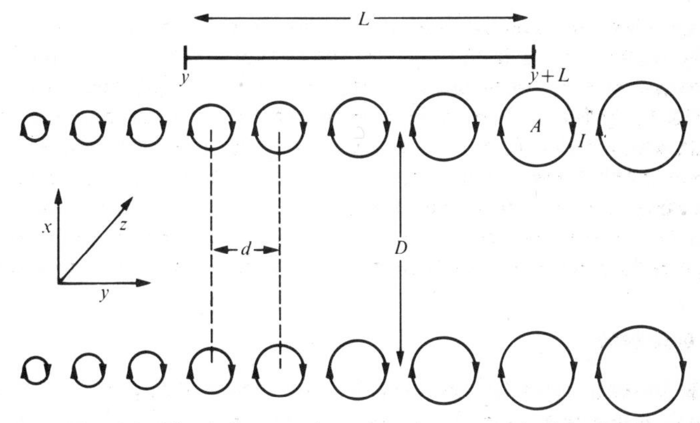

FIG. 8.2. Circulating atomic currents in a magnetized medium.

and the local dipole-moment density is $Nm = IA/Dd$ per unit area, and increases along the y-axis. The current in the x-direction crossing a macroscopic interval L extending from y to $y+L$ placed at random on the diagram is zero if either both ends of L lie outside loops or both lie inside loops. It is $+I$ if the end at $y+L$ lies within a loop, while the end at y does not, and this occurs with a probability

$$\left(\frac{A}{Dd}\right)_{y+L}\left\{1-\left(\frac{A}{Dd}\right)_{y}\right\}.$$

It is $-I$ in the converse case and this has a probability

$$\left(\frac{A}{Dd}\right)_{y}\left\{1-\left(\frac{A}{Dd}\right)_{y+L}\right\},$$

so that the expectation value is

$$\left(\frac{IA}{Dd}\right)_{y+L} - \left(\frac{IA}{Dd}\right)_y = L\frac{\partial}{\partial y}\left(\frac{IA}{Dd}\right) = L\frac{\partial}{\partial y}(Nm).$$

We can easily generalize this to three dimensions by considering planes of atoms with a separation δ parallel to the z-axis so that the interval L is replaced by an area L^2 in the yz plane and Nm becomes $IA/Dd\delta$. We then have

$$\langle I\rangle = L^2\frac{\partial}{\partial y}(Nm) \quad \text{and} \quad J_x = \frac{\langle J\rangle}{L^2} = \frac{\partial}{\partial y}(Nm).$$

If we also consider a variation of m with z and finally take other combinations of planes we eventually obtain

$$\boldsymbol{J} = \nabla \wedge (N\boldsymbol{m})$$

and this allows us to identify $N\boldsymbol{m}$ with the macroscopic vector $\boldsymbol{M}$.

The results derived in this and the last section can be given a rigorous mathematical proof using the apparatus of Fourier analysis and are exact provided only that: (a) there are an appreciable number of atoms in the smallest length which is significant on a macroscopic scale, and (2) the charge and current distributions within the atoms are adequately described in terms of their dipole moments alone. In a few cases this second condition is not fulfilled, but the only important case is in optically active media and for most purposes the identification of $N\boldsymbol{p}$ with $\boldsymbol{P}$ and $N\boldsymbol{m}$ with $\boldsymbol{M}$ is entirely adequate.

The local fields

The site of an atom is a point specified with microscopic precision and so the field acting on an individual atom may differ from the macroscopic fields $\boldsymbol{E}$ and $\boldsymbol{B}$ in a medium since, in these fields, all components which vary on an atomic scale have been averaged out. The fields may, however, still contain components $\boldsymbol{E}^{\text{loc}}$ and $\boldsymbol{B}^{\text{loc}}$ which are the same for all atoms in a macroscopic region and therefore give rise to coherent macroscopic effects. There is no general result which relates $\boldsymbol{E}^{\text{loc}}$ and $\boldsymbol{B}^{\text{loc}}$ to $\boldsymbol{E}$ and $\boldsymbol{B}$ but, in at least one important special case, we can obtain a rather simple formula.

Fig. 8.3 shows a dielectric ellipsoid in a uniform applied field $\boldsymbol{E}_0$. If the body is uniformly polarized the macroscopic field in the body is

$$\boldsymbol{E} = \boldsymbol{E}_0 - \frac{\gamma_{\text{B}}}{\epsilon_0}\boldsymbol{P}$$

where γ_{B} is the appropriate shape factor. We can however write down an exact microscopic expression for the field at the site of one atom in terms of $\boldsymbol{E}_0$ and the sum of the dipole fields of all the other atoms. If we take the origin

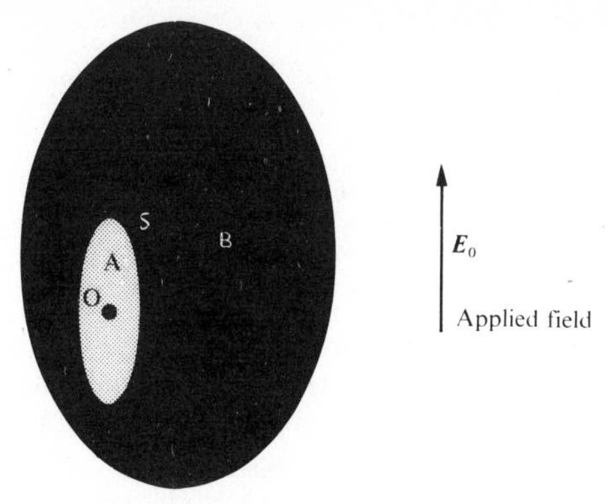

FIG. 8.3. An ellipsiodal surface S drawn within an ellipsoidal body and surrounding a point O. The region inside S is A and the region outside S is B.

to be at the atom in question the field due to an atom with a dipole moment $\boldsymbol{p}$ at $\boldsymbol{r}(n)$ is

$$\boldsymbol{e}(n) = \frac{3(\boldsymbol{p}.\boldsymbol{r}(n))\boldsymbol{r}(n)-r^2(n)\boldsymbol{p}}{4\pi\epsilon_0 r^5(n)}$$

and so

$$\boldsymbol{E}^{\text{loc}} = \boldsymbol{E}_0+\sum_n \boldsymbol{e}(n). \tag{8.2}$$

If all the atoms in the body have the same moment it can be shown that it is always possible to divide the body into two regions A and B, by an ellipsoidal surface S, around the atom at the origin, in such a way that the contribution to the sum in (8.2) from atoms in the inner region is zero. The necessary shape of S is determined by the positions of the atoms. For an isotropic medium, such as a liquid, or a cubic crystal, such as Na Cl, the surface is a sphere. We can now evaluate (8.2) simply by summing over atoms in the external region B. Since all these atoms are at a macroscopic distance from the origin this can be done as a macroscopic calculation, and clearly the field at O is $\boldsymbol{E}_0-(1/\epsilon_0)(\gamma_B-\gamma_S)\boldsymbol{P}$ where γ_S is the shape factor for the surface S. We therefore have

$$\boldsymbol{E}^{\text{loc}} = \boldsymbol{E}+\frac{\gamma_S}{\epsilon_0}\boldsymbol{P} \tag{8.3}$$

and the corresponding magnetic result is

$$\boldsymbol{B}^{\text{loc}} = \mu_0(\boldsymbol{H}+\gamma_S\boldsymbol{M}). \tag{8.4}$$

For isotropic or cubic media, where S is a sphere, $\gamma_S = \frac{1}{3}$ and we then obtain the Lorentz formulae

$$\boldsymbol{E}^{\text{loc}} = \boldsymbol{E}+\frac{1}{3\epsilon_0}\boldsymbol{P} \tag{8.5}$$

and

$$\boldsymbol{B}^{\text{loc}} = \mu_0(\boldsymbol{H}+\tfrac{1}{3}\boldsymbol{M}). \tag{8.6}$$

If the atomic polarizability is α_e we have

$$\boldsymbol{P} = N\boldsymbol{p} = N\alpha_e\epsilon_0\boldsymbol{E}^{\text{loc}} = N\alpha_e\epsilon_0\boldsymbol{E}+\gamma N\alpha_e\boldsymbol{P},$$

and so

$$\chi_e = \frac{P}{\epsilon_0 E} = \frac{N\alpha_e}{1-\gamma N\alpha_e}, \tag{8.7}$$

and also

$$\chi_m = \frac{M}{H} = \frac{N\alpha_m}{1-\gamma N\alpha_m}. \tag{8.8}$$

These calculations are only correct if the effects of all the atoms are adequately described by their dipole fields alone and all the atoms in a macroscopic region have the same moments. Even then it is possible, as for example in silicon, where the wave functions of the polarizable valence electrons extend over many atoms, that the field which is effective in polarizing the atoms is not $\boldsymbol{E}^{\text{loc}}$ at an atomic site, but nearer to $\boldsymbol{E}$ the average macroscopic field. The range of validity of the expressions (8.7) and (8.8) is therefore rather limited. The fact that the Clausius-Mossotti formula based on (8.7) (see problem 5) correctly describes the pressure and temperature dependence of the refractive index of gases shows that they are valid in this case. The success of additivity rules (see problem 6) in predicting the refractive indices of organic compounds is also evidence that they apply to molecular solids. There are few other cases where there is any compelling evidence that the Lorentz formulae (8.5) and (8.6) apply but, if we regard γ as an empirical parameter, many dielectric and magnetic media have properties which can be fitted by (8.7) and (8.8). However the uncertainty in γ makes it difficult to compare theoretical values of $N\alpha_e$ with χ_e. For example $\chi_e = 6$ is compatible with either $\gamma = \frac{1}{3}$ and $N\alpha_e = 2$ or $\gamma = \frac{1}{4}$ and $N\alpha_e = 2{\cdot}4$. If the structure of the medium is at all complex we should be unable to decide between these two values of γ.

Electric polarizability

In an electric field electrons and nuclei in atoms are displaced in opposite directions and a net dipole moment results. This process, known as electronic polarization occurs in all matter. In a crystal such as NaCl the relative displacement of the Na^+ and Cl^- ions leads to an additional effect known as ionic polarization. A somewhat similar effect occurs in molecules such as H_2O or HCl which have a permanent dipole moment even in zero field. We shall discuss this effect in connection with paramagnetism in the next section; here we consider only the induced electronic and ionic effects.

Although, in principle, these effects call for a quantum mechanical treatment the situation in any practical case is so complicated that detailed numerical results of worthwhile accuracy are almost impossible to achieve. We shall therefore only give an approximate semi-classical treatment.

Electrons are bound in atoms by the electrostatic attraction of the nucleus and the detailed configuration of the electrons is determined by a balance between this force and the dynamic properties of the electrons as described by quantum mechanics. In a field the electron configuration is distorted in such a way as to compensate the field within the atom. The distortion is therefore very like the distortion of the electron distribution which occurs in a conducting sphere in a field. Since a sphere of radius a acquires an induced moment $4\pi a^3\epsilon_0 E$ in a field E, we expect the polarizability of an atom to be related to its effective radius a by

$$\alpha_e = 4\pi a^3. \tag{8.9}$$

With N atoms in unit volume the fraction η of the space within the medium filled by atoms is $(4\pi/3)a^3 N$ and so

$$N\alpha_e \sim 3\eta. \tag{8.10}$$

In practice electronic polarizabilities with $N\alpha_e$ of between 1 and 2 are common in solids and, in gases where $\eta \sim 10^{-3}$ at S.T.P., values near 10^{-3} are found.

Ions in matter are bound to their positions of equilibrium by a complex balance of forces, but for small departures x from equilibrium, we can assume that the restoring force is $-kx$ and obeys Hooke's law. Thus, in a field E, the displacement of an ion of charge q is qE/k and the induced dipole moment is of the order of $qx = q^2E/k$. If W is the binding energy of the ion and a the equilibrium spacing we can estimate W as $\frac{1}{2}ka^2$, and so $qx \sim q^2Ea^2/2W$. Further W will be of the general order of the electrostatic binding $q^2/4\pi\epsilon_0 a$ and so we have

$$p = qx \sim \frac{q^2Ea^2}{2W} \sim 2\pi a^3\epsilon_0 E$$

and $\alpha_e \sim 2\pi a^3$. Since the number of ion pairs per unit volume is $1/2a^3$ we have

$$N\alpha_e \sim \pi \sim 3 \tag{8.11}$$

and most simple ionic media do, in fact, have ionic polarizabilities of this order.

For small displacements both ions and electrons in atoms are subject to approximately Hooke's law restoring forces and so, in an alternating field, their equations of motion are of the form

$$m\ddot{x} + kx + m\Delta\dot{x} = qE$$

where Δ is a damping term due to collisions and interactions with other particles. If we let $\omega_0 = (k/m)^{\frac{1}{2}}$ be the natural resonance frequency and set $E = E_0 \exp(j\omega t)$ the induced dipole moment, $p = p_\omega \exp(j\omega t)$, is given by

$$p_\omega = qx_\omega = \frac{\dfrac{q^2}{m}E_0}{\omega_0^2 - \omega^2 + j\omega\Delta}. \tag{8.12}$$

The static response is $p_0 = (q^2/k)E_0 = (q^2/m\omega_0^2)E_0$ and so the polarizability at ω can be expressed as

$$\alpha_e(\omega) = \alpha_e(0)\frac{\omega_0^2}{\omega_0^2 - \omega^2 + j\omega\Delta}. \tag{8.13}$$

For atomic electrons ω_0 corresponds to a line or lines in the atomic absorption spectrum and $\omega_0/2\pi$ is usually greater than 10^{15} Hz and lies in the ultra-violet. For ions the masses are larger, the restoring forces somewhat weaker and $\omega_0/2\pi$ is usually less than 10^{13} Hz and lies in the infrared.

At frequencies well below ω_0 the polarizability described by (8.13) is more or less frequency independent and equal to the static value. As ω increases towards ω_0, $N\alpha_e$ increases and becomes complex. A complex component means that $\boldsymbol{P}$ lags in phase behind $\boldsymbol{E}$ and this leads to dissipation. Waves whose frequency is anywhere near ω_0 are strongly absorbed. The absorption peak at $\omega = \omega_0$ is exceedingly intense and in general waves cannot propagate in the medium. Above ω_0, $N\alpha_e$ has a negative real part and if, as a result, the dielectric constant ϵ, which neglecting local field corrections is $1+N\alpha_e$, becomes negative, waves cannot propagate in this region either. When ω is much greater than ω_0, $N\alpha_e$ approaches zero.

At low frequencies the dielectric properties are the sum of the electronic and ionic contributions and as ω is increased the ionic term begins to dominate the behaviour until the resonance in the infrared is reached. By the time ω has been increased into the near infrared or visible ($(\omega/2\pi) > 10^{14}$ Hz) the ionic term is negligible and the dielectric properties are entirely electronic. The optical refractive index $n = \epsilon^{\frac{1}{2}}$ is therefore related solely to the electronic term. Once the resonance in the u.v. has been passed $N\alpha_e$ approaches zero and in the X-ray region all dielectric constants are virtually unity.

A local field correction gives $\chi_e = N\alpha_e/(1-\gamma N\alpha_e)$ and it is obvious that if $\gamma N\alpha_e$ approaches, or exceeds, unity the susceptibility will be very large, or there may even be a spontaneous polarization in zero field. This occurs in ferroelectric media although the detailed mechanism which leads to ferro-electricity is somewhat more complicated than this suggests. The electronic term $N\alpha_e$ is, as we have seen, of the order of 3η where η is the filling factor. At first sight η should be less than unity but the radius r which determines α_e is the radius of the electron orbits and, if these cover several atoms, r can be appreciably larger than the effective radius which determines atomic packing. Thus η can be bigger than 1 and $N\alpha_e$ bigger than 3. However, if electrons cover several atoms, they tend to see the average macroscopic field rather than $\boldsymbol{E}^{\text{loc}}$ and so the effective value of γ is small. It appears likely that $\gamma N\alpha_e$ never exceeds unity in media with no ionic contribution to $N\alpha_e$. Ferro-electricity only occurs in ionic media in which the ions also have large electronic polarizabilities.

Diamagnetism and paramagnetism

Charge of density ρ moving with a velocity $\boldsymbol{v}$ constitutes a current density $\boldsymbol{J} = \rho\boldsymbol{v}$. Thus, if we have charge circulating in an atom, we can express the magnetic moment as

$$\boldsymbol{m} = \int \tfrac{1}{2}\boldsymbol{r} \wedge \boldsymbol{J}\, \mathrm{d}V = \int \tfrac{1}{2}\boldsymbol{r} \wedge \rho\boldsymbol{v}\, \mathrm{d}V.$$

But, if the charge is carried by particles of charge q and mass M, the current is associated with a mass flow $(M/q)\rho\boldsymbol{v}$ per unit area, and the angular momentum $\boldsymbol{L}$ associated with this is

$$\boldsymbol{L} = \int \frac{M}{q}\boldsymbol{r} \wedge \rho\boldsymbol{v}\, \mathrm{d}V = \frac{2M}{q}\boldsymbol{m}.$$

The magnetic moment $\boldsymbol{m}$ is thus parallel to the angular momentum $\boldsymbol{L}$ and proportional to it. This result regarded as a relation between expectation values, is preserved in quantum mechanics, even if in addition to orbital angular momentum we also have the intrinsic spin angular momentum of the electron. Thus, if $\boldsymbol{J}$ is the total angular momentum of the atom, we can write

$$\boldsymbol{m} = g\frac{q}{2M}\boldsymbol{J} \tag{8.14}$$

where the dimensionless gyromagnetic ratio g depends on the proportions of spin and orbital angular momentum which go to make up $\boldsymbol{J}$. The angular momentum of atoms is quantized in units of $\hbar$ and so the Bohr magneton

$$\beta = \frac{e\hbar}{2M} = 9{\cdot}2741 \times 10^{-24} \text{ joule tesla}^{-1} \tag{8.15}$$

is a natural unit for atomic moments. If we express $\boldsymbol{J}$ as $\boldsymbol{j}\hbar$ where j is integral or half-integral we have

$$\boldsymbol{m} = g\beta\boldsymbol{j}. \tag{8.16}$$

A changing field B is associated with an electric field by Faraday's law and it is easy to see that if an electron is moving in a plane circular orbit of radius r normal to a field B the electric field acting round the orbit is

$$E = -\frac{1}{2\pi r}\pi r^2\dot{B} = -\tfrac{1}{2}r\dot{B}.$$

This exerts a couple $qEr = -\tfrac{1}{2}qr^2\dot{B}$ on the electron and so, by the time the field has reached its full value, the change in the angular momentum of the electron is $-\frac{1}{2}qr^2B$. The resulting change in the magnetic moment is

$$m = -\frac{1}{4}\frac{q^2}{M}r^2B, = -\frac{1}{4}\frac{e^2}{M}r^2B,$$

and so the magnetic polarizability is

$$\alpha_{\mathrm{m}} = -\frac{1}{4}\frac{e^2\mu_0 r^2}{M}.$$

To obtain an estimate of the magnitude $N\alpha_{\mathrm{m}}$ we write this as

$$N\alpha_{\mathrm{m}} = -\frac{1}{4}\frac{e^2\mu_0 r^2}{M}N = -\pi\frac{e^2}{4\pi\epsilon_0 r}\frac{\mu_0\epsilon_0}{M}Nr^3 = -\pi\frac{e^2}{4\pi\epsilon_0 r}\frac{1}{Mc^2}Nr^3.$$

Now $e^2/4\pi\epsilon_0 r$ is of the order of the electronic binding energy, about 5 eV, and Mc^2 is the rest-mass energy of the electron, about 5×10^5 eV, while πNr^3 is about equal to the atomic packing factor i.e. less than unity, thus $N\alpha_{\mathrm{m}}$ is about -10^{-5}. The resulting susceptibility is small and negative. The majority of stable ions, atoms or molecules have zero angular momentum and no permanent magnetic moment, and in this case this small negative *diamagnetic* susceptibility is the sole magnetic effect. Atoms and ions of the elements of the transition groups such as Cr, Mn, Fe etc. or the rare earths, however, have permanent moments and then the primary effect of a field is to align these moments. If the moments are free to rotate the only resistance to this alignment comes from thermal agitation associated with an energy kT. The degree of alignment is then, for small fields, proportional to mB/kT and the net moment of N atoms in the direction of the field is found to be

$$\langle Nm\rangle = \frac{m^2N}{3kT}B$$

so that

$$N\alpha_{\mathrm{m}} = \frac{\mu_0 m^2 N}{3kT}. \qquad (8.17)$$

The quantum-mechanical equivalent of this result is

$$N\alpha_{\mathrm{m}} = \frac{\mu_0 j(j+1)g^2\beta^2 N}{3kT}. \qquad (8.18)$$

The inverse dependence of $N\alpha m$ on temperature T is known as Curie's law. The values of $N\alpha_m$ which result from this formula are usually less than $1/T$ and so even this paramagnetic effect gives rise to very small values of χ_{m} except at low temperatures.

In many paramagnetic media free rotation of the magnetic ions or atoms is partially or totally restricted by interaction forces, due to their neighbouring atoms, and this somewhat modifies the simple result (8.18). Indeed in some rare-earth compounds the magnetization depends critically on the angle between the field and the crystal axes. A few media are strongly paramagnetic for one field direction and almost diamagnetic for others.

If we write $N\alpha_m = C/T$, where C is known as the Curie constant, and apply a local field correction so that

$$M = N\alpha_m H^{loc} = N\alpha_m(H+\gamma M)$$

we get

$$\chi_m = \frac{M}{H} = \frac{N\alpha_m}{1-\gamma N\alpha_m} = \frac{C}{T-\Delta}. \tag{8.19}$$

This is known as the Curie–Weiss law and clearly, as T approaches the Curie–Weiss temperature $\Delta = \gamma C$, the susceptibility becomes very large. Although (8.19) fits the experimental behaviour of many materials when T is greater than about 2Δ it is not a legitimate deduction from the local-field formula. This formula only applies if all dipoles in the medium have the same orientation and this is not true in a paramagnetic medium. In a paramagnetic medium the orientation of the dipoles which contribute to the local field at one dipole site is to some extent dependent on the orientation of the dipole itself. When this reaction term is subtracted from γM, by a procedure devised by Onsager, it leads to a form for χ_m which does not become large at any finite temperature.

The behaviour of a gas or liquid containing molecules such as HCl or H_2O which have a permanent electric dipole moment $\boldsymbol{p}$ is in some ways analogous to paramagnetism. There is no equivalent to eqn (8.18) but the classical result (8.17) is the same i.e.

$$N_S\alpha_e = \frac{p^2 N}{3\epsilon_0 kT}. \tag{8.20}$$

This is known as the Langevin–Debye Law. It predicts rather large values of $N\alpha_e$, around 10 in water at room temperature, and so if the Lorentz local-field correction formula were applicable to this case, water would be ferroelectric or spontaneously polarized. The Onsager formula, although not in very good numerical agreement with experiment, at least predicts a finite value of χ_e for water.

Although molecules with permanent electric dipole moments are described by a formula (8.20) similar to the paramagnetic law (8.17) the actual mechanism of polarization is quite different. Paramagnetic atoms are magnetized simply by there being more atoms in the medium with $\boldsymbol{m}$ parallel to $\boldsymbol{B}$ than with $\boldsymbol{m}$ opposed to $\boldsymbol{B}$. Molecules with electric dipole moments also have a quantized molecular rotation but, whereas in magnetism $\boldsymbol{m}$ is parallel to the axis of rotation, in the electric case $\boldsymbol{p}$ is normal to the axis of rotation. It turns out (see Van Vleck (1932)) that only molecules in the zero-rotational state contribute to $\boldsymbol{P}$, and their contribution is due to a polarization or deformation of the molecule. The temperature T appears in eqn (8.20) solely because it determines the number of molecules to be found in this state of zero rotational energy.

Ferromagnetism

Neither diamagnetism nor paramagnetism lead to large values of χ_m at room temperature, the large values found in materials such as iron are due to a strong *exchange interaction* between neighbouring atoms. The electrostatic interaction between nearby atoms couples the orbital motion of the electrons in the atoms and the consequences of this coupling, when combined with the requirements of the Pauli exclusion principle, can considerably lower the energy of the state where all the atoms have their magnetic moments parallel. Thus, in iron, whole macroscopic regions or domains of the medium may contain only atoms with a parallel magnetization.

As a domain grows the energy stored in its external magnetic field increases until a point is reached where the decrease in the exchange-interaction energy can no longer compensate for the increase in field energy. This balance determines the stable domain size. In iron the domain size is often limited by the size of individual crystallites which may be smaller than the stable domain size. Thus impurities and heat treatment have a large effect on the domain structure and bulk magnetic properties of iron. In an applied field domains magnetized parallel to the field grow in size at the expense of less favourably oriented domains, and also whole domains switch direction. If these processes are not inhibited by impurities, strain or grain size, the total magnetization follows the field change and we have a soft magnetic material of high susceptibility. M increases steadily with H until it saturates with $\mu_0 M \sim 2$ tesla at which point all the domains are aligned. If the domain processes are inhibited, small fields have little effect and an alignment of the domains produced by a strong field persists even in weak reverse fields. We then have a hard magnetic material suitable for permanent magnets.

Ferromagnetic properties are varied and complicated but we may note that some soft materials can have values of χ_m as large as 10^4 and so reach saturation when the intensity H is very small, e.g. $\mu_0 H \sim 10^{-4}$ tesla or in fields of 1 gauss.

Magnetic resonance

If an isolated atom of angular momentum $\boldsymbol{J}$ and magnetic moment $\boldsymbol{m} = \gamma \boldsymbol{J}$ is placed in a field $\boldsymbol{B}$ it is subject to a couple $\boldsymbol{m} \wedge \boldsymbol{B}$ and so $\mathrm{d}\boldsymbol{J}/\mathrm{d}t = \boldsymbol{m} \wedge \boldsymbol{B}$. We can write this as

$$\frac{\mathrm{d}\boldsymbol{m}}{\mathrm{d}t} = \gamma \boldsymbol{m} \wedge \boldsymbol{B}. \tag{8.21}$$

If $\boldsymbol{B}$ is a uniform field parallel to the z-axis the components of this equation are $\dot{m}_x = \gamma B m_y$, $\dot{m}_y = -\gamma B m_x$, $\dot{m}_z = 0$ and the general solution is

$m_x = A\cos(\omega t+\phi)$, $m_y = -A\sin(\omega t+\phi)$ and $m_z = C$ where $\omega = \gamma B$ and A, C, and ϕ are constants. This corresponds to $\boldsymbol{m}$ precessing about the z-axis, at a constant angle $\tan^{-1}(A/C)$ to the axis, with an angular velocity $\omega = \gamma B$. If a small oscillatory field $b\cos\omega' t$ is applied at right angles to B then when $\omega' = \omega$ energy is absorbed from this field and drives the precession, supplying the energy lost in atomic processes which damp the precession. This energy absorption can be detected by electronic means and the study of these magnetic resonance effects, and the way they are modified by the surroundings of the atom in the medium, is an important branch of solid-state physics with applications in engineering, chemistry and biophysics. For atoms the constant $\gamma = g\beta/\hbar$ is of the order of 2×10^{11} s^{-1} $tesla^{-11}$ so that in a field of 1 T the resonance occurs in the microwave region at 3×10^{10} Hz or a wavelength of 1 cm. Nuclei also possess magnetic moments associated with an intrinsic angular momentum, and nuclear magnetic resonance phenomena occur in the radio-frequency region for fields of around 1 tesla. The hydrogen nucleus, or proton, for example has a resonance at 42·57597 MHz in a field of 1 tesla. Since this constant is known with great accuracy, and it is easy to measure frequencies to six-figure accuracy, the proton magnetic-resonance frequency can be used to determine an accurate value of a magnetic field.

The frequency dependence of the magnetic susceptibility

According to eqn (8.21) the component of $\boldsymbol{m}$, of an isolated atom, parallel to an applied field is constant. Thus if a field is applied to an assembly of isolated atoms the component of the magnetization parallel to $\boldsymbol{B}$ cannot change, and so assemblies of isolated atoms cannot lead to normal magnetic effects. If a medium is to acquire a magnetization in a field it can only be as a result of interactions between the atoms which modify the equation of motion for $\boldsymbol{m}$. This is entirely different from dielectric behaviour where interactions, although they may modify the response to a field, play no such essential role.

The strength of these interactions can be described by a characteristic time τ so that if $\langle m\rangle$ is the equilibrium value of m, the effect of interactions causes m to approach $\langle m\rangle$ at a rate

$$\dot{m} = \frac{\langle m\rangle - m}{\tau}. \tag{8.22}$$

In a field of frequency ω i.e. with $B(t) = B\exp(\mathrm{j}\omega t)$, so that $\langle m\rangle$ varies as $\exp(\mathrm{j}\omega t)$, this gives

$$m = \frac{\langle m\rangle}{1+\mathrm{j}\omega\tau} \tag{8.23}$$

and so at high frequencies, where $\omega\tau$ is large, the magnetization will be small.

Values of the characteristic relaxation time τ vary enormously from medium to medium and with temperature. At low temperatures in some rare-earth compounds they may be several seconds but in every case they are longer than about 10^{-12} s. Thus at frequencies in the near infrared, and the visible, the magnetization can no longer follow the field. For this reason magnetic properties have no influence on optical properties and it is usual in discussing the propagation of electromagnetic waves of optical frequencies to assume that all media have zero magnetic susceptibility and $\mu = 1$.

PROBLEMS

8.1. Calculate the mean distance between the centres of the molecules in water, and the wavelength of yellow light in water.

8.2. The exact charge density in a region of space is given by

$$\rho(x, y, z) = \rho_0 \cos\frac{2\pi x}{a}\cos\frac{2\pi y}{a}\cos\frac{2\pi z}{a}$$

where a is a microscopic length. What is the macroscopic charge density at x, y, z?

8.3. A uniform line of equal dipoles p with a regular spacing d extends from $x = 0$ to large values of x. Consider a macroscopic interval of length L, one of whose ends is outside the array at $x < 0$ and the other in the array at $x+L > 0$, and show that the effective macroscopic charge in L is p/d. How is this calculation relevant to the notion of an effective charge at the surface of a polarized dielectric?

8.4. A dielectric has a uniform polarization $+\mathrm{P}_z$ and extends from $z = 0$ to $z = Z$. What is the sign of the charge of the sub-atomic particle nearest to the surface at Z?

8.5. The refractive index n of a medium is related to χ_e by $n^2 = 1+\chi_e$. Use eqns (8.1a) and (8.5) to show that

$$\frac{n^2-1}{n^2+2} = \frac{\chi_e}{\chi_e+3} = \tfrac{1}{3}N\alpha_e.$$

This is known as the Clausius–Mossotti formula. If α_e for the molecules in a gas is independent of pressure and temperature what does this result predict about the variation of n with pressure and temperature?

8.6. The molar refractivity R of an atomic structural unit such as, for example, the electrons associated with a C—H bond in a hydrocarbon is defined by $R = \frac{1}{3}L\alpha_e$ where L is Avagadro's number and α_e the polarizability of the bond. If the Lorentz formula (8.5) is correct, show that the refractive index n of a compound in which in unit volume there are k_1 moles of a unit of refractivity R_1, k_2 of a unit of refractivity R_2, etc. can be expressed as

$$\frac{n^2-1}{n^2+2} = k_1R_1+k_2R_2+\text{etc.}$$

8.7. Show that when a conducting sphere of radius a is placed in a uniform electric field $\boldsymbol{E}$ it acquires an induced dipole moment $4\pi a^3\epsilon_0\boldsymbol{E}$.

8.8. Sketch the real and imaginary parts of $N\alpha_e(\omega)$ given by eqn (8.13) as a function of ω/ω_0. Assume that $\Delta = 0{\cdot}1\omega_0$.

8.9. Why are many colourless transparent materials opaque for infrared radiation?

8.10. Why is the refractive index of a medium never greater than the square root of its static dielectric constant?

8.11. Water is diamagnetic and $\chi_m = -9\times10^{-6}$. The Mn^{2+} ion is magnetic with $j = \frac{5}{2}$ and $g = 2$. Use eqn (8.18) to calculate the mass of $MnCl_2$ (molecular weight 126) that must be added to 1 litre of water at $T = 300$ K to make χ_m exactly zero. (The Bohr magneton $\beta = 9{\cdot}274\times10^{-24}$ JT^{-1}, Boltzmann's constant $k = 1{\cdot}381\times10^{-23}$ J K^{-1}, and Avogadro's number is $6{\cdot}022\times10^{23}$ per mole.)

8.12. Why cannot X-rays be focused by a lens?

8.13. On the assumption that each iron atom has a magnetic moment of 1 Bohr magneton estimate the saturation magnetization of iron. The molecular weight of iron is 56 and its relative density is 8.

8.14. Sodium chloride has a refractive index $n = 1{\cdot}5$ and a static dielectric constant $\epsilon = 6$. The Na^+ and Cl^- ions are located at alternate corners of cubes of side 4×10^{-10} m and have charges of $\pm e$. Calculate the displacement of the ions in a static electric field of 10^6 Vm^{-1} parallel to a cube edge.

8.15. An ionic crystal has a static dielectric constant $\epsilon(0)$ and the square of its refractive index is $n^2 = \epsilon(\infty)$. The ionic resonance occurs at $\omega_0/2\pi = 10^{13}$ Hz. Show that in the vicinity of this resonance the dielectric constant can be expressed as $\epsilon(\omega) = \epsilon(\infty)+(\epsilon(0)-\epsilon(\infty))(1-\omega^2/\omega_0^2)^{-1}$. If $\epsilon(\infty) = 2{\cdot}5$ and $\epsilon(0) = 5$ find the range of frequencies over which $\epsilon(\omega)$ is negative.

8.16. In c.g.s. units (commonly used in the classic texts of atomic and solid state physics) $B = H$ in vacuum, and both M and H have the same units as B (gauss) but $B = H+4\pi M$. The Bohr magneton is then expressed in ergs gauss^{-1} and Boltzmann's constant k in ergs degree^{-1}. Verify that the c.g.s. equivalent of the Curie law formula (8.18) is obtained simply by omitting μ_0.

9. Energy and power

Introduction

THE rate $\phi\dot{q}$ at which energy is supplied to charge a plane parallel capacitor of area A and plate separation d can be expressed as $EdA\dot{D}$ and this suggests that the field energy density is increasing at a rate $E\dot{D}$. The magnetic intensity in a long solenoid of length l and radius r with n turns per unit length is related to the current I by $H = nI$ and the back e.m.f. is $\phi = \pi r^2 nl\dot{B}$. Thus the rate $I\phi$ at which energy is supplied can be expressed as $\pi r^2 lH\dot{B}$ and this suggests that the magnetic-field energy density is increasing at a rate $H\dot{B}$. We now consider these results in more general terms.

The force per unit volume acting on charge and current is $\boldsymbol{F} = \rho\boldsymbol{E}+\boldsymbol{J} \wedge \boldsymbol{B}$ and, if the current is due to the motion of charged particles with a velocity $\boldsymbol{v}$, we have $\boldsymbol{J} = \rho\boldsymbol{v}$. The rate at which this force does work on the charges is $\boldsymbol{v}.\boldsymbol{F} = \rho\boldsymbol{v}.\boldsymbol{E}+\rho\boldsymbol{v}.(\boldsymbol{v} \wedge \boldsymbol{B}) = \rho\boldsymbol{v}.\boldsymbol{E} = \boldsymbol{J}.\boldsymbol{E}$. The magnetic force does no work since it is at right angles to the motion. Now the fields satisfy $\nabla \wedge \boldsymbol{E} = -\dot{\boldsymbol{B}}$ and $\nabla \wedge \boldsymbol{H} = \dot{\boldsymbol{D}}+\boldsymbol{J}$ and so

$$\nabla.(\boldsymbol{E} \wedge \boldsymbol{H}) = \boldsymbol{H}.(\nabla \wedge \boldsymbol{E})-\boldsymbol{E}.(\nabla \wedge \boldsymbol{H}) = -(\boldsymbol{H}.\dot{\boldsymbol{B}}+\boldsymbol{E}.\dot{\boldsymbol{D}}+\boldsymbol{E}.\boldsymbol{J}).$$

If we integrate each term over a closed volume V with a surface S, whose positive normal is outwards, the volume integral of $\nabla.(\boldsymbol{E} \wedge \boldsymbol{H})$ becomes a surface integral and we obtain

$$\oint_S (\boldsymbol{E} \wedge \boldsymbol{H}).\mathrm{d}\boldsymbol{S}+\int_V \{\boldsymbol{E}.\boldsymbol{J}+\boldsymbol{E}.\dot{\boldsymbol{D}}+\boldsymbol{H}.\dot{\boldsymbol{B}}\}\,\mathrm{d}V = 0. \tag{9.1}$$

The second term represents the rate at which work is being done on the charged particles in V and thus either going to increase their kinetic energy or being dissipated as heat when the particles make collisions. It is therefore plausible that the other two volume integrals represent the rate of increase of field energy and that the surface integral gives the rate at which energy flows outwards across S. The equation would then be a statement of the law of conservation of energy. We now justify this interpretation.

If the surface S is a closed conducting metal wall then $\boldsymbol{E}$ has no component tangential to S and $\boldsymbol{E} \wedge \boldsymbol{H}$ no component normal to S. The surface integral is then zero. If inside S all the currents are steady so that $\dot{\boldsymbol{B}} = 0$, then, since $\int \boldsymbol{E}.\boldsymbol{J}\,\mathrm{d}V$ is the rate at which the non-electromagnetic energy is increasing, $-\int \boldsymbol{E}.\dot{\boldsymbol{D}}\,\mathrm{d}V$ must be the rate of decrease of the potential energy associated with the field. If the currents are not steady the term $-\int \boldsymbol{H}.\dot{\boldsymbol{B}}\,\mathrm{d}V$ must represent an extra term in the rate of decrease of field energy. With these terms identified we can replace the physical wall at S by a purely mathematical surface and identify the surface integral as the outward energy flux. The vector

$$\boldsymbol{N} = \boldsymbol{E} \wedge \boldsymbol{H} \tag{9.2}$$

is known as Poynting's vector. It is the energy-flux vector of the electromagnetic field and gives the power transmitted across unit area by the fields. Problems 9.2 and 9.3 illustrate its use in two familiar contexts but its real importance is that it can be used in dealing with systems of waves where simpler methods fail.

In a system containing only dielectric media, the energy associated with a change in the field can be expressed as

$$\delta U = \int \boldsymbol{E}.\delta\boldsymbol{D}\,\mathrm{d}V \tag{9.3}$$

and this must be supplied by external sources e.g. batteries. Suppose then that the field is due to a set of electrodes with charges q_k at potentials ϕ_k completely enclosed in a conducting screen at a potential $\phi_0 = 0$. We shall have

$$\int \boldsymbol{E}.\boldsymbol{D}\,\mathrm{d}V = -\int \boldsymbol{D}.\nabla\phi\,\mathrm{d}V = -\int\{\nabla.(\phi\boldsymbol{D}) - \phi\nabla.\boldsymbol{D}\}\,\mathrm{d}V = -\int \nabla.(\phi\boldsymbol{D})\,\mathrm{d}V$$

since $\nabla.\boldsymbol{D} = 0$. The volume integral of the divergence gives a surface integral which is a sum of terms, one for each conducting electrode. In each of these terms ϕ_k is constant on the surface and so

$$\int \boldsymbol{E}.\boldsymbol{D}\,\mathrm{d}V = -\sum_k \phi_k \int_k \boldsymbol{D}.\mathrm{d}\mathbf{S}_k.$$

Since the outward normal of the system is the inward normal of the electrodes these surface integrals are just the electrode charges and

$$\int \boldsymbol{E}.\boldsymbol{D}\,\mathrm{d}V = \sum_k \phi_k q_k. \tag{9.4}$$

For a small change δq_k in the electrode charges

$$\int \boldsymbol{E}.\delta\boldsymbol{D}\,\mathrm{d}V = \sum_k \phi_k\,\delta q_k, \tag{9.5}$$

which confirms our identification of $\boldsymbol{E}.\dot{\boldsymbol{D}}$ with the rate of change of electric-field energy density. We now use these results to discuss the work done to place a dielectric body in an applied field and the force acting on it.

Work and energy in a dielectric system

An applied field $\boldsymbol{E}^*$ is set up in a region of empty space by a set of electrodes with charges q_k^* and potentials ϕ_k^*. (We are using the notation of Landau and Lifshitz (1960).) The work required to do this is

$$U_{\mathrm{mt}} = \iint_0^E \boldsymbol{E}^* . \epsilon_0 \, \mathrm{d}\boldsymbol{E}^* \, \mathrm{d}V = \frac{1}{2} \int \epsilon_0 \boldsymbol{E}^* . \boldsymbol{E}^* \, \mathrm{d}V. \tag{9.6}$$

The electrodes are isolated so that the charges q_k^* remain constant and a dielectric body is inserted into the field. This changes the field to $\boldsymbol{E}$, and also $\boldsymbol{D}$ is no longer equal to $\epsilon_0 \boldsymbol{E}$ within the body. The potentials of the electrodes also change to ϕ_k although the charges $q_k = q_k^*$ are unchanged. The energy U of this system is the sum of U_{mt}, the energy U_0 of the body outside the field, and the mechanical work W required to place the body in the field. We can choose the origin of energy so that $U_0 = 0$, and then $W = U - U_{\mathrm{mt}}$. We cannot calculate U directly but we can calculate the change in U when the electrode charges are changed. Thus for a small charge $\delta q_k = \delta q_k^*$ in the electrode charges the work required to place the body in the field is increased by

$$\delta W = \delta U - \delta U_{\mathrm{mt}} = \int (\boldsymbol{E} . \delta \boldsymbol{D} - \epsilon_0 \boldsymbol{E}^* . \delta \boldsymbol{E}^*) \, \mathrm{d}V. \tag{9.7}$$

To simplify this expression we note that

$$\boldsymbol{E} . \delta \boldsymbol{D} - \epsilon_0 \boldsymbol{E}^* . \delta \boldsymbol{E}^* = (\boldsymbol{E} . \delta \boldsymbol{D} - \boldsymbol{E} . \delta \epsilon_0 \boldsymbol{E}^*) + (\epsilon_0 \boldsymbol{E} - \boldsymbol{D}) . \delta \boldsymbol{E}^* + (\boldsymbol{D} - \epsilon_0 \boldsymbol{E}^*) . \delta \boldsymbol{E}^*$$

and that the volume integrals of the first and last pairs of terms are

$$\sum_k \phi_k (\delta q_k - \delta q_k^*) \quad \text{and} \quad \sum_k (q_k - q_k^*) \, \delta \phi_k^*.$$

These are both zero since $\delta q_k = \delta q_k^*$ and $q_k = q_k^*$ because the charges are the same with and without the body present. In the remaining term $\boldsymbol{D} - \epsilon_0 \boldsymbol{E}$ is zero, except in the body where it is equal to $\boldsymbol{P}$, and so

$$\delta W = - \int_{\text{body}} \boldsymbol{P} . \delta \boldsymbol{E}^* \, \mathrm{d}V. \tag{9.8}$$

If the applied field is uniform or varies slowly over the dielectric body (note that $\boldsymbol{E}^*$ is not the field in the body, that is $\boldsymbol{E}$) we can replace (9.8) by

$$\delta W = -\boldsymbol{p} . \delta \boldsymbol{E}^* \tag{9.9}$$

where $\boldsymbol{p} = \int \boldsymbol{P} \, \mathrm{d}V$ is the total dipole moment of the body. If $\boldsymbol{p}$ is independent of $\boldsymbol{E}^*$ the work required to place the body in the field is $-\boldsymbol{p} . \boldsymbol{E}^*$ but if $\boldsymbol{p}$ is proportional to $\boldsymbol{E}^*$ it is $-\frac{1}{2}\boldsymbol{p} . \boldsymbol{E}^*$.

If a body is placed in a region where the applied field is $\boldsymbol{E}^*$ and then moved through $\mathrm{d}\mathbf{r}$ to a region where the field is $\boldsymbol{E}^*+\mathrm{d}\boldsymbol{E}^*$, the mechanical work required is $-\boldsymbol{p}.\mathrm{d}\boldsymbol{E}^*$ and so the force $\boldsymbol{F}$ acting on the body satisfies $-\boldsymbol{F}.\mathrm{d}\mathbf{r} = -\boldsymbol{p}.\mathrm{d}\boldsymbol{E}^*$. The x-component is therefore

$$F_x = p_x \frac{\partial E_x^*}{\partial x} + p_y \frac{\partial E_y^*}{\partial y} + p_z \frac{\partial E_z^*}{\partial z}. \qquad (9.10)$$

If in addition, $\boldsymbol{p}$ is proportional to $\boldsymbol{E}^*$ so that $\boldsymbol{p} = \epsilon_0 \alpha_e \boldsymbol{E}^*$, we have

$$F_x = \tfrac{1}{2}\epsilon_0 \alpha_e \frac{\partial}{\partial x}\{E_x^{*2} + E_y^{*2} + E_z^{*2}\}$$

or

$$\boldsymbol{F} = \tfrac{1}{2}\epsilon_0 \alpha_e \nabla(E^{*2}). \qquad (9.11)$$

We can use this to calculate either the force on a polarizable atom or particle, or the force on a volume $\mathrm{d}V$ of a polarizable fluid. In this case α_e is replaced by $\chi_e\,\mathrm{d}V$ and the force per unit volume is

$$\boldsymbol{f} = \tfrac{1}{2}\epsilon_0 \chi_e \nabla(E^{*2}). \qquad (9.12)$$

It attracts the fluid into regions where the field is stronger.

If the applied field is produced by two electrodes, one of them earthed, the additional work required to introduce a body into the space between the electrodes when there is an additional charge δq^* on the live electrode is $-\boldsymbol{p}.\delta\boldsymbol{E}^*$. This must be the additional energy required to add the charge δq^* with the body in place and, if the potential is ϕ with the body in place and ϕ^* in its absence, we have

$$(\phi - \phi^*)\,\delta q^* = -\boldsymbol{p}.\delta\boldsymbol{E}^*.$$

Now for two parallel plates of area A we have $\delta q^* = \epsilon_0 A\,\delta E^*$, and so

$$\phi - \phi^* = -\frac{p_n}{\epsilon_0 A} \qquad (9.13)$$

where p_n is the component of the dipole moment normal to the plates. If we give the plates a constant charge and measure the change in potential as the body is introduced we can measure p_n, whatever the shape of the body.

Work in magnetic systems

We can perform similar calculations for a magnetic body introduced into a magnetic field $\boldsymbol{B}^* = \mu_0 \boldsymbol{H}^*$. The calculation is a little more involved since in this case we have to consider the currents which produce $\boldsymbol{B}^*$ as held constant while the body is introduced. The result is, however, the same. If the body has a moment $\boldsymbol{m}$ the additional work required to introduce it into a field augmented by $\delta \boldsymbol{B}^*$ is

$$\delta W = -\boldsymbol{m}.\delta \boldsymbol{B}^*. \tag{9.14}$$

We may perhaps remark that many texts—reference (8) is a notable exception—are misleading on this point.

Thermodynamics

The two expressions $-\boldsymbol{m}.\delta \boldsymbol{B}^*$ and $-\boldsymbol{p}.\delta \boldsymbol{E}^*$ refer to energy supplied to a system as work. If energy is also supplied as heat δQ the change in the total energy is, in the magnetic case,

$$\delta U^* = \delta Q - m\,\delta B^* \tag{9.15}$$

where for brevity we have dropped vector notation. If the change occurs reversibly at a temperature T we can express δQ in terms of an entropy change as $T\,\delta S$ and so $\delta U^* = T\,\delta S - m\,\delta B^*$. The function $F^* = U^* - TS$ is the difference between the free energy of the system and the free energy of the system less the body. It is not the true free energy of the system, that is $F = U - TS$, but it is still a useful thermodynamic function of the variables T and B^* which define the state of the system. We have

$$\delta F^* = \delta U^* - T\,\delta S - S\,\delta T = -S\,\delta T - m\,\delta B^*, \tag{9.16}$$

and so

$$S = -\left(\frac{\partial F^*}{\partial T}\right)_{B^*} \quad \text{and} \quad m = -\left(\frac{\partial F^*}{\partial B^*}\right)_T. \tag{9.17}$$

We can then derive the Maxwell relation

$$\left(\frac{\partial S}{\partial B^*}\right)_T = \left(\frac{\partial m}{\partial T}\right)_{B^*}. \tag{9.18}$$

If a body has a susceptibility which decreases with increasing temperature then $(\partial m/\partial T)_{B^*}$ is negative and so $T(\partial S/\partial B^*)_T$ is also negative, and heat is evolved as a field is applied. This is the basic result used in magnetic cooling experiments at low temperatures.

Although F^* is not the true thermodynamic free energy it is the free-energy function obtained from a statistical mechanical calculation and so eqns (9.17) are also important in this context.

Dissipation in dielectric media

According to the results of the first section the rate at which external sources supply energy to unit volume of a dielectric is $\boldsymbol{E}.\dot{\boldsymbol{D}}$. If $\boldsymbol{E}$ and $\boldsymbol{D}$ vary with time so that $\boldsymbol{E}(t) = \text{Re}\,\boldsymbol{E}_0 \exp(\text{j}\omega t)$ and $\boldsymbol{D}(t) = \text{Re}\,\boldsymbol{D}_0 \exp(\text{j}\omega t)$, the time average of $\boldsymbol{E}.\dot{\boldsymbol{D}}$ is $\frac{1}{2}\,\text{Re}(\boldsymbol{E}_0^*.\text{j}\omega\boldsymbol{D}_0)$ and this must be the rate at which energy is dissipated in unit volume. If we express $\boldsymbol{D}_0$ in terms of $\boldsymbol{E}_0$ as $\boldsymbol{D}_0 = (\epsilon' - \text{j}\epsilon'')\epsilon_0\boldsymbol{E}_0$ the complex dielectric constant corresponds to a phase lag δ where $\tan\delta = \epsilon''/\epsilon'$ between $\boldsymbol{D}(t)$ and $\boldsymbol{E}(t)$ and we have, for the energy dissipated,

$$Q = \tfrac{1}{2}\omega\epsilon_0\epsilon''\boldsymbol{E}_0^*.\boldsymbol{E}_0 = \tfrac{1}{2}\omega\epsilon_0\epsilon'\tan\delta\,\boldsymbol{E}_0^*.\boldsymbol{E}_0. \tag{9.19}$$

A complex dielectric constant or susceptibility, therefore, is associated with dissipation.

PROBLEMS

9.1. Verify that $\nabla.(\boldsymbol{E}\wedge\boldsymbol{H}) = \boldsymbol{H}.(\nabla\wedge\boldsymbol{E})-\boldsymbol{E}.(\nabla\wedge\boldsymbol{H})$.

9.2. A current I flows in a round wire of radius r and resistance R per unit length. Find the values of $\boldsymbol{E}$ and $\boldsymbol{H}$ at the surface of the wire. Show that $\boldsymbol{E}\wedge\boldsymbol{H}$ is directed inwards towards the wire. Integrate $\boldsymbol{E}\wedge\boldsymbol{H}$ over the surface of unit length of wire and interpret your result.

9.3. A coaxial cable carries a current I and the inner conductor of radius a is at a steady potential V relative to the outer conductor, whose radius is b. Calculate $\boldsymbol{E}$, $\boldsymbol{H}$ and $\boldsymbol{E}\wedge\boldsymbol{H}$ in the region between the conductors and integrate $\boldsymbol{E}\wedge\boldsymbol{H}$ over the cross-section of the cable.

9.4. In a plane electromagnetic wave in vacuum $\boldsymbol{E}$ and $\boldsymbol{H}$ are perpendicular to each other and the direction of propagation, and $E = (\mu_0/\epsilon_0)^{\frac{1}{2}}H$. Estimate the electric field at the earth's surface due to a satellite radio-transmitter at a height of 100 km radiating 10 watts equally distributed in all directions.

9.5. One end of a vertical U tube containing a solution of density 1 g cm^{-3} and susceptibility $\chi_{\mathrm{m}} = 10^{-3}$ is in a magnetic field of 1 tesla, and the other end is outside the field. What is the difference in height between the liquid in the two limbs of the tube?

9.6. Calculate the induced dipole moment of a sphere of radius a and dielectric constant ϵ in a uniform applied field $\boldsymbol{E}$. If the sphere is placed between two parallel plates of area A and separation d calculate the resulting change in the capacitance between the plates.

9.7. A liquid has a dielectric constant $\epsilon = 3+(1\,000/T)$ at a temperature T. How much heat is evolved per unit volume when an applied electric field E is slowly raised to 10^{-6} V m^{-1}?

9.8. Obtain the magnetic equation equivalent to eqn (9.19).

9.9. An electric heating element runs at 1 000 K and according to Stefan's law the power radiated per m^2 at a temperature T is $6(T/100)^4$ watts m^{-2}. Estimate the r.m.s. value of the electric field at the surface of the element.

9.10. Calculate the force acting on an iron atom of moment 1 Bohr magneton in a region where the field is 1 tesla and the field gradient 10^4 T m^{-1}. What is the resultant acceleration?

10. Electromagnetic waves

Introduction

WAVES of one kind or another occur in almost every branch of physics, and although perhaps the most easily visualized waves are those that may exist on the surface of a liquid, the simplest waves are the current and voltage waves which can be excited in a cable or transmission line. These waves exhibit in the most elementary way features which are common to all other forms of wave motion. They involve two parameters, the current I and voltage V, coupled together by a pair of first-order partial-differential equations. The product of these variables represents energy flow and both variables are intrinsically continuous quantities. We therefore begin this chapter with a discussion of these waves. The remainder of the chapter, except for the section on guided waves, deals with plane electromagnetic waves in homogeneous media. The way in which currents and charges generate electromagnetic waves is dealt with in the next chapter.

Transmission lines

Any two-wire cable or line has a definite inductance L and capacitance C per unit length and, as a result, in a length $\mathrm{d}z$ the voltage between the conductors drops by $\dot{I}L\,\mathrm{d}z$ and the current I is reduced by $\dot{V}C\,\mathrm{d}z$ due to leakage across the stray capacitance. With the sign convention illustrated in Fig. 10.1 the voltage and current obey the equations

$$\frac{\partial V}{\partial z} = -L\dot{I} \text{ (a)} \quad \text{and} \quad \frac{\partial I}{\partial z} = -C\dot{V}. \text{ (b)} \tag{10.1}$$

A function of the form $f(u)$ where $u = \omega t - \beta z$ represents a forward wave propagating with a velocity ω/β and a function $r(v)$ where $v = \omega t + \beta z$

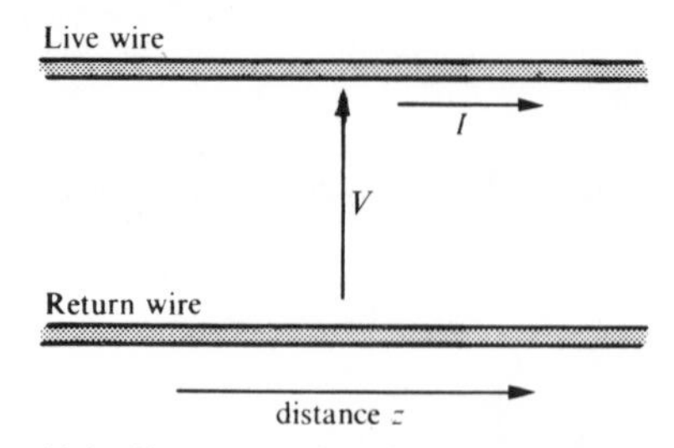

FIG. 10.1. Sign conventions for a transmission line.

represents a reverse wave. It is easy to verify that, if we express I as $I = I_f f(u) + I_r r(v)$ and V as $V_f f(u) + V_r r(v)$, where I_f, $I_r V_f$ and V_r are constants, then these functions satisfy the pair of eqns (10.1) if

$$\frac{\omega}{\beta} = (LC)^{-\frac{1}{2}} \tag{10.2}$$

and

$$V_f = Z_0 I_f, \qquad V_r = -Z_0 I_r, \tag{10.3}$$

where

$$Z_0 = \left(\frac{L}{C}\right)^{\frac{1}{2}}. \tag{10.4}$$

Thus the cable or line can support either forward or reverse waves propagating with a velocity $(LC)^{-\frac{1}{2}}$ and, in the forward wave, the voltage is in phase with the current, whereas in the reverse wave it is out of phase. With the sign convention of Fig. 10.1 the power flowing to the right is

$$IV = (I_f f(u) + I_r r(v)) Z_0 (I_f f(u) - I_r r(v)) = Z_0 (I_f^2 f^2(u) - I_r^2 r(v))$$

and we see that the sign reversal in (10.3) is connected with the way the forward wave carries power forwards and the reverse wave carries power backwards.

A special and important type of wave is the harmonic or sinusoidal wave of frequency $\omega/2\pi$ and wavelength $\lambda = 2\pi/\beta$. A forward wave might be described in real terms by $I_{f0}\cos(\omega t - \beta z + \phi)$ but we shall find it more convenient to use complex notation $I_f \exp\{j(\omega t - \beta z)\}$. These two expressions are equivalent if we remember that we have eventually to take the real part of the complex expression and also set $I_f = I_{f0}\exp(j\phi)$. Because $\exp(j\omega t)$ is common to all expressions we can omit it and, for example, express forward waves as $I_f \exp(-j\beta z)$ with $V_f = Z_0 I_f$. We note that in complex notation the average power flow is $\frac{1}{2}\,\mathrm{Re}\, IV^*$ and this leads to

$$P = \tfrac{1}{2} Z_0 (I_f I_f^* - I_r I_r^*) = \frac{1}{2Z_0}(V_f V_f^* - V_r V_r^*). \tag{10.5}$$

Clearly β and ω are related by $\beta = \omega(LC)^{\frac{1}{2}}$, but in real lines the conductors will have a finite resistance R per unit length and so V and I satisfy the equations

$$\frac{\partial V}{\partial z} = -(L\dot{I} + RI) \tag{10.6a}$$

and

$$\frac{\partial I}{\partial z} = -C\dot{V} \tag{10.6b}$$

and these lead to the relation

$$-\beta^2 = (-j\omega L + R)(-j\omega C)$$

or

$$\beta = \omega(LC)^{\frac{1}{2}}\left(1+\frac{R}{\mathrm{j}\omega L}\right). \tag{10.7}$$

If R is small we have

$$\beta = \beta'-\mathrm{j}\beta'' = \omega(LC)^{\frac{1}{2}}-\frac{\mathrm{j}}{2}\frac{R}{Z_0}, \tag{10.8}$$

and β is complex. As a result the wave amplitude decays with distance as $\exp(-\beta''z)$ and the power or intensity as $\exp(-2\beta''z)$. Ohmic loss leads to a complex propagation constant β and to attenuation.

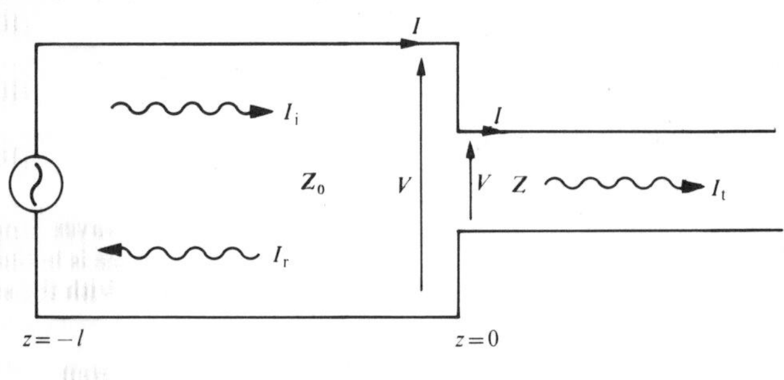

FIG. 10.2. A junction between two transmission lines of different characteristic impedances, Z_0 and Z.

We now look at the significance of the quantity Z_0 which is known as the characteristic impedance of the line. In Fig. 10.2 we show a line of impedance Z_0 connected at $z = 0$ to a line of impedance Z, which is so long that its natural attenuation will eliminate all reflections from its far end. The signal generator will launch an incident wave of amplitude I_i in the line of impedance Z_0 and, at the junction, the current and voltage in this line are $I = I_i+I_r$ and $V = V_i+V_r = Z_0(I_i-I_r)$, where I_r corresponds to a reflected wave. In the second line we have only a forward or transmitted wave and the current and voltage are $I = I_t$ and $V = V_t = ZI_t$. But I and V must be continuous at the junction and so we have $I_i+I_r = I_t$ and $Z_0(I_i-I_r) = ZI_t = Z(I_i+I_r)$. The amplitude of the reflected wave is thus determined by Z_0 and Z as

$$\frac{I_r}{I_i} = \frac{Z_0-Z}{Z_0+Z} = -\frac{V_r}{V_i} \tag{10.9}$$

and the fraction of the incident power reflected at the junction is therefore

$$r^2 = \frac{\frac{1}{2}Z_0I_rI_r^*}{\frac{1}{2}Z_0I_iI_i^*} = \left|\frac{Z_0-Z}{Z_0+Z}\right|^2. \tag{10.10}$$

Unless $Z = Z_0$ some fraction of the power will be reflected back to the generator at $z = -l$. We should have obtained an exactly similar result if, instead of a long line of impedance Z, we had connected a circuit of impedance Z to the first line at $z = 0$. Thus, if a line of impedance Z_0 is connected to a load Z, it will be correctly terminated, or matched, so that there are no reflections only if $Z = Z_0$.

The current in the line at the generator end at $z = -l$ is $I = I_f \exp(j\beta l) + I_r \exp(-j\beta l)$ and the voltage is $V = Z_0\{I_f \exp(j\beta l) - I_r \exp(-j\beta l)\}$. The ratio $Z_{in} = V/I$ is the input impedance presented by the line to the generator. If we use (10.9) to eliminate I_r in terms of I_i we obtain, after some rearrangement

$$Z_{in} = Z_0 \frac{Z \cos \beta l + jZ_0 \sin \beta l}{Z_0 \cos \beta l + jZ \sin \beta l}. \tag{10.11}$$

Clearly if the line is short and the frequency low, so that $\beta l = \omega l(LC)^{\frac{1}{2}}$ is small, we obtain $Z_{in} \sim Z$, which is the usual low-frequency approximation. If, however, βl is appreciable Z_{in} may be quite different from Z. In particular if $\beta l = \pi/2$ which corresponds to a line with $l = \lambda/4$, we get $Z_{in} = Z_0^2/Z$ and Z_{in} is inversely proportional to Z. A quarter-wave line can be used to produce an impedance transformation. If for example we wish to connect a line of impedance Z_1 to a line of impedance Z_2 we can eliminate reflections at one frequency by inserting a quarter-wave section of line of impedance $Z_0 = (Z_1 Z_2)^{\frac{1}{2}}$.

It can be shown that for any line in vacuum consisting of straight parallel conductors, e.g. a coaxial line or a parallel wire line, $LC = \mu_0\epsilon_0$ and so the wave velocity is $(LC)^{-\frac{1}{2}} = (\mu_0\epsilon_0)^{-\frac{1}{2}} = c$ the velocity of light. If the space between or around the conductors is filled with a medium of dielectric constant ϵ this increases C, but leaves L unchanged, and so the velocity is reduced to $c' = c/\epsilon^{\frac{1}{2}}$ and the impedance Z_0 is also reduced by the same factor. In a lossy dielectric $\boldsymbol{D}$ lags in phase behind $\boldsymbol{E}$ and, if δ is the phase angle at a particular frequency, we can describe this by a complex dielectric constant

$$\epsilon = \epsilon' - j\epsilon'' = \epsilon'(1 - j \tan \delta).$$

We then find that $\beta = \omega(\mu_0\epsilon_0\epsilon)^{\frac{1}{2}}(1 - j \tan \delta)^{\frac{1}{2}}$ and, if $\tan \delta$ is small,

$$\beta = \beta' - j\beta'' = \omega(\mu_0\epsilon'\epsilon_0)^{\frac{1}{2}}(1 - \tfrac{1}{2}j \tan \delta).$$

Again there will be attenuation and the intensity of a wave decays as $\exp(-2\beta''z)$. We can express this in terms of the wavelength $\lambda' = 2\pi/\beta'$ as $\exp\{-(2\pi \tan \delta/\lambda')z\}$. If, for example $\tan \delta = 10^{-3}$ and $\lambda' = 10$ cm $= 0{\cdot}1$ m the intensity falls by a factor $\exp(-20\pi) \sim 10^{-27}$ in 1 km. This clearly has rather serious implications for the use of transmission lines at microwave frequencies.

Plane waves in homogeneous, isotropic media

In a homogeneous, isotropic medium $\boldsymbol{B} = \mu\mu_0\boldsymbol{H}$ and $\boldsymbol{D} = \epsilon\epsilon_0\boldsymbol{E}$ are parallel to $\boldsymbol{H}$ and $\boldsymbol{E}$ and μ and ϵ do not vary with position. The equations $\nabla.\boldsymbol{B} = 0$ and $\nabla.\boldsymbol{D} = 0$ then imply $\nabla.\boldsymbol{H} = 0$ and $\nabla.\boldsymbol{E} = 0$. There is, as we shall see, a considerable advantage in working in terms of the vectors $\boldsymbol{E}$ and $\boldsymbol{H}$ rather than any other pair and so we shall write the field equations as

$$\nabla.\boldsymbol{H} = 0 \text{ (10.12a)}, \qquad \nabla.\boldsymbol{E} = 0 \text{ (10.12b)}$$

$$\nabla \wedge \boldsymbol{E} = -\mu\mu_0\dot{\boldsymbol{H}} \text{ (10.12c)} \quad \text{and} \quad \nabla \wedge \boldsymbol{H} = \epsilon\epsilon_0\dot{\boldsymbol{E}} \text{ (10.12d)}.$$

From (10.12c) and (10.12d) we obtain

$$\nabla \wedge (\nabla \wedge \boldsymbol{E}) = -\mu\mu_0 \nabla \wedge \dot{\boldsymbol{H}} = -\mu\mu_0\epsilon\epsilon_0\ddot{\boldsymbol{E}}.$$

Now if E_x is a Cartesian component of $\boldsymbol{E}$ we have

$$[\nabla \wedge (\nabla \wedge \boldsymbol{E})]_x = \frac{\partial}{\partial x}(\nabla.\boldsymbol{E}) - \nabla^2 E_x,$$

and so, since $\nabla.\boldsymbol{E} = 0$ we obtain

$$\nabla^2 E_x = \mu\mu_0\epsilon\epsilon_0\ddot{E}_x. \tag{10.13}$$

There are similar equations for the other components of $\boldsymbol{E}$, and also by operating on (10.12d) and (10.12c) in the reverse order we obtain

$$\nabla^2 H_x = \mu\mu_0\epsilon\epsilon_0\ddot{H}_x \tag{10.13b}$$

with similar equations for H_y and H_z. Eqn (10.13a) is the three-dimensional wave equation and, amongst its solutions, there are functions which correspond to plane waves. Thus if $E_x = f(u)$ where $u = \omega t - \boldsymbol{\beta}.\boldsymbol{r}$ we have $\nabla^2 E_x = \beta^2(\mathrm{d}^2E_x/\mathrm{d}u^2)$ and $\ddot{E}_x = \omega^2(\mathrm{d}^2E_x/\mathrm{d}u^2)$ so that E_x satisfies (10.13a) if $\beta = \omega(\mu\mu_0\epsilon\epsilon_0)^{\frac{1}{2}}$. A function such as $f(u)$ represents a wave propagating with a velocity ω/β in a direction parallel to $\boldsymbol{\beta}$ and since, at a fixed time, f is constant over planes for which $\boldsymbol{\beta}.\boldsymbol{r} = 0$ the wave fronts are planes normal to the direction of propagation.

We now look at these plane-wave solutions in more detail and to do this we need to go back to the field equations. We assume that we have plane waves in which $\boldsymbol{E} = \boldsymbol{E}_0 \exp\{\mathrm{j}(\omega t - \boldsymbol{\beta}.\boldsymbol{r})\}$ and $\boldsymbol{H} = \boldsymbol{H}_0 \exp\{\mathrm{j}(\omega t - \boldsymbol{\beta}.\boldsymbol{r})$ where $\boldsymbol{E}_0$ and $\boldsymbol{H}_0$ are constant vectors, so that these waves have a frequency $\omega/2\pi$ and a wavelength $\lambda = 2\pi/\beta$. We also note that for these waves

$$\nabla.\boldsymbol{H} = -\mathrm{j}\boldsymbol{\beta}.\boldsymbol{H}_0 \exp\{\mathrm{j}(\omega t - \boldsymbol{\beta}.\boldsymbol{r})\} = -\mathrm{j}\boldsymbol{\beta}.\boldsymbol{H}, \tag{10.14a}$$

$$\nabla \wedge \boldsymbol{E} = -\mathrm{j}\boldsymbol{\beta} \wedge \boldsymbol{E}_0 \exp\{\mathrm{j}(\omega t - \boldsymbol{\beta}.\boldsymbol{r})\} = -\mathrm{j}\boldsymbol{\beta} \wedge \boldsymbol{E}, \tag{10.14b}$$

and of course $\dot{\boldsymbol{E}} = \mathrm{j}\omega\boldsymbol{E}$. The two divergence equations now yield $\boldsymbol{\beta}.\boldsymbol{H} = 0$ and $\boldsymbol{\beta}.\boldsymbol{E} = 0$ and so, if plane-wave solutions of the field equations exist, they have no field components in the direction of propagation. Eqn (10.12d) yields $\omega\epsilon\epsilon_0\boldsymbol{E} = -\boldsymbol{\beta} \wedge \boldsymbol{H}$ and so $\boldsymbol{E}$ and $\boldsymbol{H}$ are not only perpendicular to $\boldsymbol{\beta}$ but also to each other. Eqn (10.12c) yields $\omega\mu\mu_0\boldsymbol{H} = \boldsymbol{\beta} \wedge \boldsymbol{E}$ and so we have $\omega^2\mu\mu_0\epsilon\epsilon_0\boldsymbol{E} = -\boldsymbol{\beta} \wedge (\boldsymbol{\beta} \wedge \boldsymbol{E})$. But $\boldsymbol{\beta} \wedge (\boldsymbol{\beta} \wedge \boldsymbol{E}) = \boldsymbol{\beta}(\boldsymbol{\beta}.\boldsymbol{E}) - \beta^2\boldsymbol{E} = -\beta^2\boldsymbol{E}$ since $\boldsymbol{\beta}.\boldsymbol{E} = 0$, and so finally

$$\beta = \omega(\mu\mu_0\epsilon\epsilon_0)^{\frac{1}{2}} \tag{10.15}$$

and the waves propagate with a velocity

$$c' = (\mu\mu_0\epsilon\epsilon_0)^{-\frac{1}{2}}. \tag{10.16}$$

In vacuum, where $\epsilon = \mu = 1$, this reduces to the velocity of light

$$c = (\mu_0\epsilon_0)^{-\frac{1}{2}} = 2{\cdot}997295 \times 10^8 \text{ m s}^{-1}. \tag{10.17}$$

Thus there are plane-wave solutions of the field equations and they are purely transverse waves which propagate with the velocity of light in vacuum. At optical frequencies where $\mu = 1$ the refractive index of a dielectric medium is

$$n = c/c' = \epsilon^{\frac{1}{2}}, \tag{10.18}$$

but we must remember that the appropriate value of ϵ will only contain electronic contributions and so may be very different from the static or low-frequency value, which can also contain an ionic term.

If we write $\boldsymbol{\beta}$ as $\beta\boldsymbol{u}$, so that $\boldsymbol{u}$ is a unit vector in the direction of propagation, we can express the relation between $\boldsymbol{E}$ and $\boldsymbol{H}$ as

$$\boldsymbol{E} = \frac{\beta}{\omega\epsilon\epsilon_0}\boldsymbol{H} \wedge \boldsymbol{u} = Z\boldsymbol{H} \wedge \boldsymbol{u}, \tag{10.19}$$

where

$$Z = \left(\frac{\mu\mu_0}{\epsilon\epsilon_0}\right)^{\frac{1}{2}} = \left(\frac{\mu}{\epsilon}\right)^{\frac{1}{2}} Z_0. \tag{10.20}$$

The quantity Z is the impedance of the medium and Z_0 the impedance of free space is approximately 120π ohms. Note that if $\mu = 1$ we have $Z = Z_0/n$. This impedance relates $\boldsymbol{E}$ to $\boldsymbol{H}$ and plays much the same role as the characteristic impedance of a transmission line which relates V to I. We note also that Poynting's vector is $\boldsymbol{N} = \boldsymbol{E} \wedge \boldsymbol{H} = ZH^2\boldsymbol{u}$ or, if we remember that we are using complex notation, the time average of $\boldsymbol{N}$ is

$$\tfrac{1}{2}Z(\boldsymbol{H}_0.\boldsymbol{H}_0^*)\boldsymbol{u} = \frac{1}{2Z}(\boldsymbol{E}_0.\boldsymbol{E}_0^*)\boldsymbol{u}.$$

This not only verifies that energy propagates with the wave but also allows us to relate the intensity to the field amplitudes.

If we take the direction of propagation as the z-axis, the wave will contain two independent pairs of components $E_x = ZH_y$ and $E_y = -ZH_x$, and these correspond to the independent linearly polarized components of a light wave in optics. If the wave propagates in the negative z-direction we obtain $E_x = -ZH_y$ and $E_y = ZH_x$ and the sign reversal in these relations corresponds to a sign reversal in the energy flux N_z. Consider now a plane wave propagating parallel to the z-axis in a medium with a dielectric constant ϵ_1 and with $\mu = 1$ and $Z_1 = Z_0/\epsilon_1^{\frac{1}{2}}$, which falls at normal incidence on the plane surface of a medium with constants ϵ_2, $\mu = 1$ and $Z_2 = Z_0/\epsilon_2^{\frac{1}{2}}$. We assume that the wave is polarized with the single pair of field components E_x and H_y. These components are tangential to the surface between the two media and must be continuous. Thus if E_i, H_i correspond to the incident wave and $E_t H_t$ to the transmitted wave there will also have to be a reflected wave with components E_r, H_r in the first medium, to ensure that $E_i + E_r = E_t$ and $H_i + H_r = H_t$ while at the same time $E_i = Z_1 H_i$, $E_r = -Z_1 H_r$ and $E_t = Z_2 H_t$. Thus we obtain

$$H_i + H_r = H_t \quad \text{and} \quad Z_1(H_i - H_r) = Z_2 H_t$$

so that

$$\frac{H_r}{H_i} = \frac{Z_1 - Z_2}{Z_1 + Z_2} = -\frac{E_r}{E_i}. \tag{10.21}$$

The fraction of the incident intensity reflected is

$$r^2 = \frac{H_r H_r^*}{H_i H_i^*} = \left|\frac{Z_1 - Z_2}{Z_1 + Z_2}\right|^2 \tag{10.22}$$

which we can also express in terms of the refractive indices $n_1 = \epsilon_1^{\frac{1}{2}}$ and $n_2 = \epsilon_2^{\frac{1}{2}}$ as

$$r^2 = \left(\frac{n_1 - n_2}{n_1 + n_2}\right)^2. \tag{10.23}$$

This is Fresnel's formula in optics. We leave it to the reader to check that exactly similar formulae result if the wave is polarized with field components E_y and H_x.

The analogy between the behaviour of waves at normal incidence and waves in a transmission line can be taken further. A quarter-wave layer of a medium of impedance Z_3 placed on a medium of impedance Z_2 transforms the wave impedance to Z_3^2/Z_2. This is the principle used in blooming lenses and other optical components to minimize reflections. Thus a layer of a medium of refractive index $n^{\frac{1}{2}}$ evaporated on to a glass surface of refractive index n will eliminate reflections at the frequency at which the layer is a quarter of a wavelength thick. In optical instruments this frequency is usually chosen to correspond to green light and, as a result, white light produces predominantly red and blue reflections, which give the surface a characteristic purple appearance.

Refraction and reflection

When a plane wave falls, at any angle other than normal incidence, on the surface of a medium with a different dielectric constant, conditions at the interface become rather more complicated; but the nature of the reflected and transmitted waves is still determined by the requirement that the tangential components of $\boldsymbol{E}$ and $\boldsymbol{H}$ be everywhere continuous. In Fig. 10.3 a

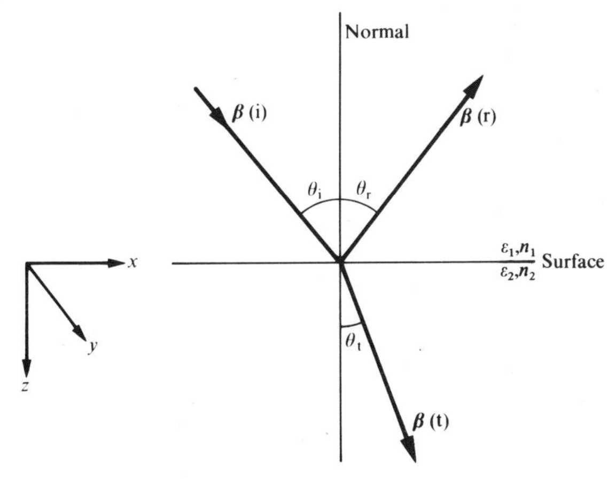

FIG. 10.3. Plane waves incident, reflected, and transmitted at a plane dielectric interface.

plane wave propagating in the xz plane as $\exp\{j(\omega t - \boldsymbol{\beta}(i).\boldsymbol{r})\}$ is incident on a plane dielectric interface in the xy plane at $z = 0$. The components of $\boldsymbol{\beta}(i)$ are $\beta_x = \beta(i)\sin\theta_i$, $\beta_y = 0$ and $\beta_z = \beta(i)\cos\theta_i$, where θ_i is the angle of incidence. On the dielectric interface the fields associated with this wave vary as $\exp(-j\beta_x x)$ and these fields, together with the fields associated with the reflected wave, must match the fields associated with the transmitted wave at all points on the interface. Referring to the figure we see that this implies $\beta(i)\sin\theta_i = \beta(r)\sin\theta_r = \beta(t)\sin\theta_t$. Since the incident and reflected waves propagate in the same medium $\beta(i) = \beta(r) = \omega(\mu_0\epsilon_0\epsilon_1)^{\frac{1}{2}}$ and so the angles of incidence and reflection $\theta_i = \theta_r$ are the same. On the other hand

$$\beta(t) = \left(\frac{\epsilon_2}{\epsilon_1}\right)^{\frac{1}{2}}\beta(i) = \frac{n_2}{n_1}\beta(i),$$

and so $\sin\theta_t = (n_1/n_2)\sin\theta_i$ which is Snell's law. Furthermore, since β_y must be zero for all three waves, the directions of incidence, reflection and refraction lie in a plane which includes the normal. So far we have merely reproduced the results of geometrical optics. We now move on to look at the magnitude of the reflection coefficient. This depends on the polarization of the incident

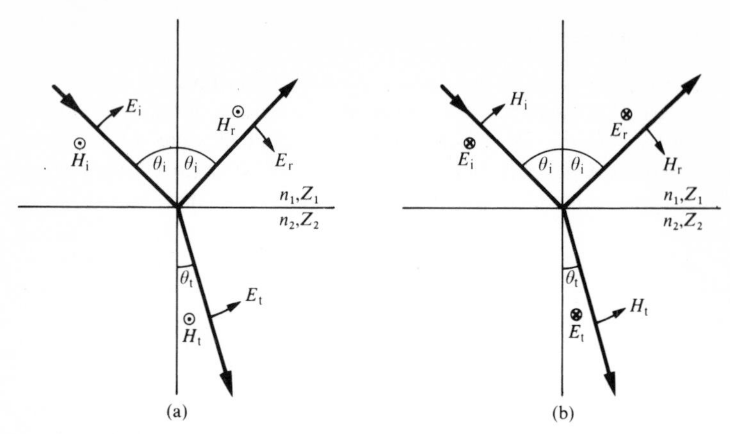

FIG. 10.4. Fields associated with waves of two different polarizations incident on a plane dielectric interface.

light. If the light is polarized as in Fig. 10.4(a) the tangential components of $\boldsymbol{H}$ are simply H_i, H_r and H_t but the tangential components of $\boldsymbol{E}$ are $E_i \cos \theta_i$, $E_r \cos \theta_r$ and $E_t \cos \theta_t$. Thus for the fields to match at the surface we require $H_i + H_r = H_t$ and $(E_i + E_r)\cos \theta_i = E_t \cos \theta_t$ or

$$Z_1(H_i - H_r)\cos \theta_i = Z_2 H_t \cos \theta_t,$$

and this leads to a reflection coefficient

$$r^2 = \frac{H_r^2}{H_i^2} = \left(\frac{Z_1 \cos \theta_i - Z_2 \cos \theta_t}{Z_1 \cos \theta_i + Z_2 \cos \theta_t}\right)^2 = \left(\frac{n_2 \cos \theta_i - n_1 \cos \theta_t}{n_2 \cos \theta_i + n_1 \cos \theta_t}\right)^2. \quad (10.24)$$

For the polarization shown in Fig. 10.4(b) we have $E_i + E_r = E_t$, but $(H_i + H_r)\cos \theta_i = H_t \cos \theta_t$ and this leads to

$$r^2 = \left(\frac{n_1 \cos \theta_i - n_2 \cos \theta_t}{n_1 \cos \theta_i + n_2 \cos \theta_t}\right)^2. \quad (10.25)$$

Now if $n_2 > n_1$ Snell's law gives $\theta_i > \theta_t$ and so $\cos \theta_t > \cos \theta_i$. It is therefore possible for the numerator of (10.24) to vanish but not the numerator of (10.25). The angle of incidence at which this occurs satisfies

$$\frac{n_2}{n_1} \cos \theta_i = \cos \theta_t \quad \text{and also} \quad \frac{n_1}{n_2} \sin \theta_i = \sin \theta_t,$$

so that $(n_2/n_1)^2 \cos^2\theta_i + (n_1/n_2)^2 \sin^2\theta_i = 1$ which gives $\tan \theta_i = n_2/n_1$. It is known as Brewster's angle. The windows in a gas-laser tube are usually set at Brewster's angle to the axis of the tube in order to minimize internal reflections.

Waves in conductors. The skin effect

In a metal of conductivity σ there is a conduction-current density $\boldsymbol{J} = \sigma\boldsymbol{E}$ in addition to the displacement-current density $\epsilon_0\dot{\boldsymbol{E}}$ and so the field equation $\nabla \wedge \boldsymbol{H} = \epsilon_0\dot{\boldsymbol{E}} = \mathrm{j}\omega\epsilon_0\boldsymbol{E}$ is replaced by $\nabla \wedge \boldsymbol{H} = \boldsymbol{J}+\epsilon_0\dot{\boldsymbol{E}} = (\sigma+\mathrm{j}\omega\epsilon_0)\boldsymbol{E}$. In most metals σ is greater than 10^6 mho m^{-1} and so, since $\epsilon_0 \sim 10^{-11}$ Fm^{-1} the conduction current $\boldsymbol{J}$ is much greater than the displacement current $\epsilon_0\dot{\boldsymbol{E}}$ at all normal frequencies. In other words $\omega\epsilon_0 \ll \sigma$ and we have approximately $\nabla \wedge \boldsymbol{H} = \sigma\boldsymbol{E}$ instead of $\nabla \wedge \boldsymbol{H} = \mathrm{j}\omega\epsilon_0\boldsymbol{E}$. This is only modification to the field equations and therefore the propagation constant $\beta = \omega(\mu\mu_0\epsilon_0)^{\frac{1}{2}}$ is replaced by

$$\beta = \left(\frac{\omega\mu\mu_0\sigma}{\mathrm{j}}\right)^{\frac{1}{2}} = \left(\frac{\omega\mu\mu_0\sigma}{2}\right)^{\frac{1}{2}}(1-\mathrm{j}) = \frac{1-\mathrm{j}}{\delta}, \tag{10.26}$$

and the impedance $Z = (\mu\mu_0/\epsilon_0)^{\frac{1}{2}}$ by

$$Z = \left(\frac{\mathrm{j}\omega\mu\mu_0}{\sigma}\right)^{\frac{1}{2}} = \left(\frac{\omega\mu\mu_0}{2\sigma}\right)^{\frac{1}{2}}(1+\mathrm{j}). \tag{10.27}$$

We see from (10.25) that β is complex and a wave propagating in a conductor decreases in amplitude with distance as $\exp(-z/\delta)$. The characteristic length δ is known as the skin depth. In copper it is about 1 cm at 50 Hz but in iron, because μ is large, it is less than 1 mm. Since δ varies as $\omega^{-\frac{1}{2}}$ it becomes very small at high radio-frequencies and even smaller at optical frequencies.

Fig. 10.5 shows a thick block of metal in which a current of density J flows in response to a tangential electric field E_0 at the surface. The current density at the surface is σE_0 and at a depth z we have

$$J = \sigma E = \sigma E_0 \exp\left\{-\mathrm{j}\left(\frac{1-\mathrm{j}}{\delta}\right)z\right\} = \sigma E_0 \exp\left\{-\left(\frac{1+\mathrm{j}}{\delta}\right)z\right\}.$$

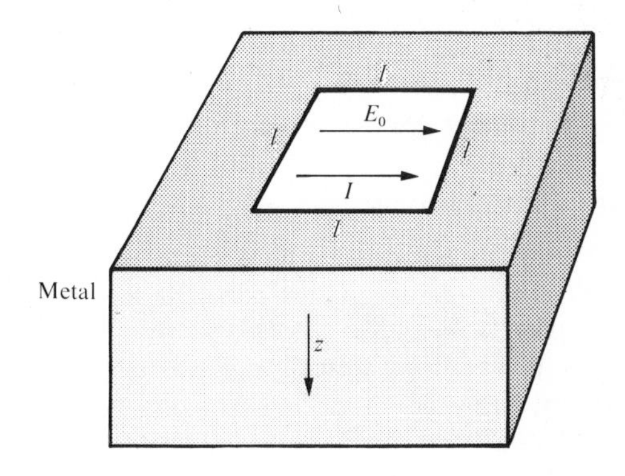

FIG. 10.5. A square element of area on the surface of a metal block.

The total current flowing under the square of side l is therefore

$$I = l\sigma E_0 \int_0^\infty \exp\left\{-\left(\frac{1+\mathrm{j}}{\delta}\right)z\right\} \mathrm{d}z = \frac{l\sigma\delta}{1+\mathrm{j}} E_0.$$

The voltage difference between opposite sides of the square is lE_0, and so there is apparently a surface impedance

$$Z_s = R_s + \mathrm{j}X_s = \frac{1+\mathrm{j}}{\sigma\delta} \text{ ohms.} \tag{10.27}$$

We note that this is equal to the impedance Z given by (10.26). We can use (10.27) to calculate the impedance per unit length of a round wire of radius r much greater than δ and, since the periphery of the wire is $2\pi r$, it is $Z_S/2\pi r$. The resistive part $R = 1/2\pi r\sigma\delta$ may be compared with the d.c. resistance $R_0 = 1/\pi r^2\sigma$ and it is greater by a factor $r/2\delta$ which may be quite large.

If a plane wave falls at normal incidence from vacuum on a metal surface the reflection coefficient is simply

$$r^2 = \left|\frac{Z_0 - Z}{Z_0 + Z}\right|^2$$

where $Z_0 = (\mu_0/\epsilon_0)^{\frac{1}{2}}$ and Z is given by (10.26). This leads, if $\mu = 1$, to

$$r^2 = \left|\frac{1 - \left(\dfrac{\omega\epsilon_0}{2\sigma}\right)^{\frac{1}{2}}(1+\mathrm{j})}{1 + \left(\dfrac{\omega\epsilon_0}{2\sigma}\right)^{\frac{1}{2}}(1+\mathrm{j})}\right|^2$$

and, since $(\omega\epsilon_0/2\sigma)^{\frac{1}{2}}$ is small, to

$$r^2 \sim 1 - 4\left(\frac{\omega\epsilon_0}{2\sigma}\right)^{\frac{1}{2}}.$$

Even at optical frequencies this is near unity for most metals.

Electromagnetic waves and momentum

The mean energy flux associated with a plane wave is $\boldsymbol{N} = \frac{1}{2}(\boldsymbol{E} \wedge \boldsymbol{H}^*)$ where for brevity we have omitted the 'real part of' sign. If the wave propagates in an almost empty region containing a small concentration of charge carriers, such as electrons, which gives it a small conductivity σ but does not appreciably alter the nature of the wave, the energy dissipated as heat per unit volume is $\frac{1}{2}\sigma\boldsymbol{E}.\boldsymbol{E}^*$ which we can also express as the magnitude of

$$\tfrac{1}{2}\sigma\left(\frac{\mu_0}{\epsilon_0}\right)^{\frac{1}{2}}\boldsymbol{E} \wedge \boldsymbol{H}^*.$$

Thus in a distance dz the change in $\boldsymbol{N}$ is d$\boldsymbol{N} = -\sigma(\mu_0/\epsilon_0)^{\frac{1}{2}}\boldsymbol{N}\,\mathrm{d}z$. The current density is $\boldsymbol{J} = \sigma\boldsymbol{E}$ and the magnetic force per unit volume acting on the current is $\mu_0\boldsymbol{J} \wedge \boldsymbol{H}$, which has a time average value $\frac{1}{2}\mu_0\sigma\boldsymbol{E} \wedge \boldsymbol{H}^*$. The force per unit area acting on a thin lamina of thickness dz, as the wave passes through it, is therefore $\mathrm{d}\boldsymbol{F} = \mu_0\sigma\boldsymbol{N}\,\mathrm{d}z = -(\mu_0\epsilon_0)^{\frac{1}{2}}\,\mathrm{d}\boldsymbol{N} = -(1/c)\,\mathrm{d}\boldsymbol{N}$. Thus as the energy flux decreases by $-\mathrm{d}\boldsymbol{N}$ momentum is transferred to the medium at a rate $\mathrm{d}\boldsymbol{F} = -(1/c)\,\mathrm{d}\boldsymbol{N}$ per unit area. This momentum must have come from the wave and we conclude that the momentum flux associated with the wave is

$$\boldsymbol{\pi} = \frac{1}{c}\boldsymbol{N}. \tag{10.28}$$

This result is quite general for plane waves in vacuum, and leads to a number of important conclusions. If photons of energy $\hbar\omega$ cross unit area at a rate n per second the energy flux is $n\hbar\omega$ and the momentum flux is $n\hbar\omega/c = n\hbar\beta$. Thus each photon has a momentum $\hbar\beta$. This result is the basis from which de Broglie developed wave mechanics. If radiation of intensity $\boldsymbol{N}$ falls at normal incidence on a reflecting surface, the momentum of the radiation is reversed on reflection and a force per unit area, or pressure, equal to $2N/c$ acts on the surface. This is just equal to the sum u of the energy densities associated with the incident and reflected waves and so the pressure is $p = u$. Waves incident at an angle θ to the normal have a momentum component $\pi\cos\theta$ normal to the surface and approach the surface at a rate $c\cos\theta$, thus in this case $p = u\cos^2\theta$. For thermal radiation incident at all angles, an average over θ yields $p = \frac{1}{3}u$, and this formula is the basis for the thermodynamic treatment of radiation given by Stefan and Wien. It was Planck's investigation of the connection between Wien's results and Rayleigh and Jean's statistical radiation law that led him to the discovery of quantum theory.

Waves in plasmas

A plasma is a neutral ionized gas containing free electrons and positive ions. In a high-frequency electric field the light electrons respond to the field, whereas the more massive ions remain almost fixed in position. The equation of motion of an electron is

$$\ddot{\boldsymbol{r}} = \frac{q}{m}\boldsymbol{E}\exp(\mathrm{j}\omega t) \quad \text{and so} \quad \boldsymbol{r} = -\frac{q}{m\omega^2}\boldsymbol{E}\exp(\mathrm{j}\omega t).$$

With N electrons in unit volume the polarization is $\boldsymbol{P} = Nq\boldsymbol{r}$ and therefore the dielectric susceptibility is $\chi = -Nq^2/\epsilon_0 m\omega^2 = -\omega_\mathrm{p}^2/\omega^2$. The quantity $\omega_\mathrm{p}/2\pi$ is known as the plasma frequency. Values of $\omega_\mathrm{p}/2\pi$ between about 10^6 and 10^{12} Hz occur in laboratory plasmas and, in the ionosphere, $\omega_\mathrm{p}/2\pi$ is of the order of 20 MHz. The electrons and lattice ions in metals and semiconductors also behave in some ways like a plasma and in metals $\omega_\mathrm{p}/2\pi$ is of the order of 10^{15} Hz whereas in semiconductors it may be anywhere between about 10^8 and 10^{12} Hz depending on the free-carrier concentration. The dielectric constant is

$$\epsilon = 1+\chi = 1-\frac{\omega_p^2}{\omega^2} \tag{10.29}$$

and so the propagation constant β of a wave satisfies

$$\beta^2 = \omega^2\mu_0\epsilon_0\left(1-\frac{\omega_p^2}{\omega^2}\right) = \omega^2\mu_0\epsilon_0-\omega_p^2\mu_0\epsilon_0. \tag{10.30}$$

If $\omega > \omega_p$, $\epsilon < 1$ and β is less than $\omega(\mu_0\epsilon_0)^{\frac{1}{2}}$: thus the phase velocity of the wave ω/β is greater than c. However, from (10.30), we have

$$\frac{\omega}{\beta}\frac{d\omega}{d\beta} = \frac{d(\omega^2)}{d(\beta^2)} = c^2, \tag{10.31}$$

and so the group velocity $d\omega/d\beta$, i.e. the velocity with which a pulse of radiation traverses the medium is less than c. If $\omega < \omega_p$ we find that $\epsilon < 0$ and β is imaginary so that no wave propagation occurs. The plasma frequency essentially marks the boundary between a low-frequency region in which conduction processes dominate and a high-frequency region in which dielectric behaviour is observed. In semiconductor devices ω_p is the ultimate upper limit to high-frequency operation.

If $\omega < \omega_p$ the impedance $Z = (\mu_0/\epsilon\epsilon_0)^{\frac{1}{2}}$ is purely imaginary, say $Z = jX$, and, at the surface of the medium an incident wave is reflected. The reflection coefficient is

$$r^2 = \left(\frac{Z_0-Z}{Z_0+Z}\right)^2 = \left|\frac{Z_0-jX}{Z_0+jX}\right|^2 = 1.$$

Thus the reflection is complete. This occurs for low-frequency radio-waves at the beginning of the ionosphere and explains why long-distance radio-communication is possible. Higher-frequency radio-waves penetrate the ionosphere and are lost, so that only line of sight communication is possible. We may give a somewhat similar explanation of why metals reflect light, infrared and radio-waves but not X-rays. In practical cases these simple results are appreciably modified by the effects of collisions between the charged particles of the plasma.

The dielectric constant ϵ can also be negative in a dielectric medium just above an electronic or ionic resonance and in this region β and Z are again imaginary. Thus waves can neither penetrate nor propagate in dielectrics in these frequency ranges, which tend to occur in the ultraviolet and infrared regions of the spectrum.

Guided waves

A wave propagating within or near a material structure is known as a guided wave. The wave in a coaxial cable is a simple example of a guided

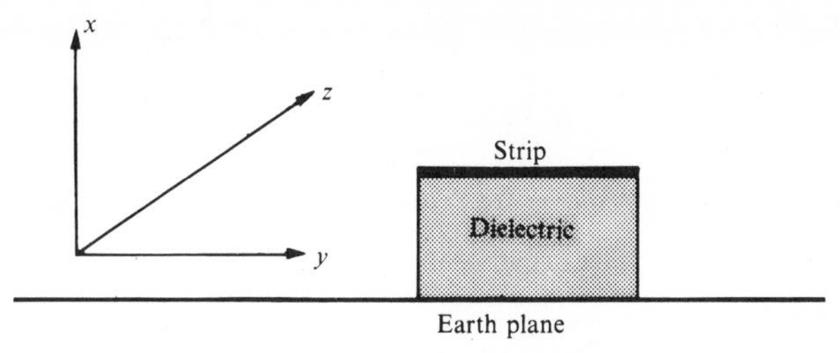

FIG. 10.6. Cross-section of a microwave strip-line.

wave. Another simple and familiar example is a light wave propagating in a light pipe where it is totally internally reflected at the walls. A somewhat less familiar example is the microwave strip-line shown in cross-section in Fig. 10.6. This consists of a conducting strip separated from a conducting earth plane by a low-loss medium of high dielectric constant ϵ. A wave propagates along the strip and the propagation constant, $\beta \sim \omega(\mu_0 \epsilon \epsilon_0)^{\frac{1}{2}}$, is greater than ω/c. The fringing fields at the edge of the strip excite fields outside the strip-line but these fields must satisfy the wave equation in vacuum. Thus, if they vary with x and y as $\exp(-jax)$ and $\exp(-jby)$, we must have

$$a^2+b^2+\beta^2 = \frac{\omega^2}{c^2}$$

and, since $\beta^2 > \omega^2/c^2$, a and b must be imaginary. In other words the fringing field will decay exponentially to zero at a distance from the line. The wave is therefore trapped in the region near the line and is guided by the line. This illustrates a very general result. Any system that propagates a wave with a phase velocity ω/β less than c can act as a waveguide.

The commonest form of waveguide is, however, rather different and consists of a metal pipe, usually of rectangular cross-section, and the wave propagates within the pipe. The system is shown in cross-section in Fig. 10.7. We do not propose to go into a detailed discussion of propagation in such a guide but merely to show the existence of a guided wave and some of its properties. The tangential component of $\boldsymbol{E}$ must vanish at the walls of the

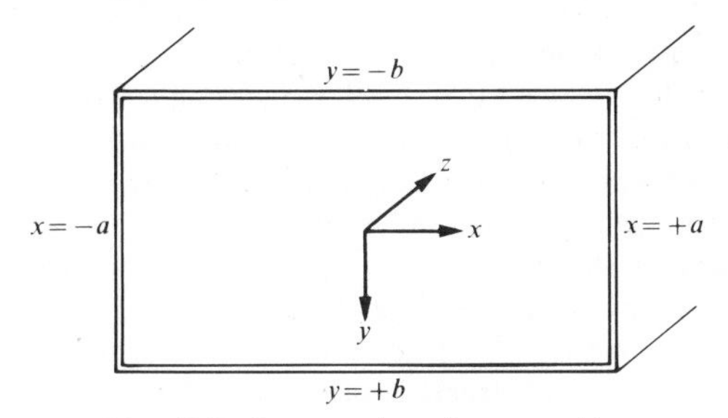

FIG. 10.7. Cross-section of a waveguide.

guide and it is easy to see that this will be the case if $\boldsymbol{E}$ has a single component

$$E_y = E_0 \cos \frac{\pi x}{2a} \exp\{j(\omega t - \beta z)\}. \tag{10.32}$$

We can obtain the components of $\boldsymbol{H}$ associated with this wave from

$$-j\omega\mu_0 \boldsymbol{H} = -\mu_0 \dot{\boldsymbol{H}} = \nabla \wedge \boldsymbol{E},$$

and they are

$$H_x = -\frac{\beta}{\omega\mu_0} E_0 \cos \frac{\pi x}{2a} \exp\{j(\omega t - \beta z)\} \tag{10.33}$$

$$H_z = \frac{\pi}{2j\omega\mu_0 a} E_0 \sin \frac{\pi x}{2a} \exp\{j(\omega t - \beta z)\}. \tag{10.34}$$

We see also that the time average of Poynting's vector has a single non-vanishing component

$$\langle N_z \rangle = -\tfrac{1}{2} \operatorname{Re} E_y^* H_x = +\frac{1}{2}\frac{\beta}{\omega\mu_0} E_0^2 \cos^2 \frac{\pi x}{2a} \tag{10.35}$$

and so the wave transmits power down the guide. The wave has, however, a component of $\boldsymbol{H}$ in the direction of propagation and this is permissible because it is not a plane wave. Each field component must satisfy the three-dimensional wave equation and $\nabla^2 E_y = \mu_0\epsilon_0 \dot{E}_y$ yields the relation

$$\beta^2 = \omega^2\mu_0\epsilon_0 - \left(\frac{\pi}{2a}\right)^2 = \left(\frac{2\pi}{\lambda_0}\right)^2 - \left(\frac{\pi}{2a}\right)^2 \tag{10.36}$$

where λ_0 is the free-space wavelength corresponding to ω. We see that this has a real solution for β only if $\lambda_0 < 4a$. Thus there can be no wave propagation unless the transverse dimension of the guide $2a$ exceeds $\lambda_0/2$. The value $\lambda_c = 4a$ is known as the cut-off wavelength. We also note that, when $\lambda_0 < \lambda_c$, the phase velocity ω/β is greater than c. However, from (10.36), we have

$$\frac{\omega}{\beta}\frac{d\omega}{d\beta} = \frac{d(\omega^2)}{d(\beta^2)} = c^2, \tag{10.37}$$

and so the group velocity $d\omega/d\beta$ is always less than c. As λ_0 approaches λ_c, the phase velocity becomes infinite and the group velocity zero. We note also that, from (10.33), $H_x \to 0$ while H_z remains finite. At cut-off the waves bounce fruitlessly between the walls instead of propagating along the guide.

Similar guided waves propagate in conducting tubes of any cross-section and, whatever the cross-section, the propagation constant β is related to ω by an equation of the form $\beta^2 = \omega^2\mu_0\epsilon_0 - (2\pi/\lambda_c)^2$ in which the cut-off wavelength λ_c is fixed by the shape and dimensions of the cross-section. Thus the relation (10.37) between the phase and group velocities is quite general. Guided waves are of considerable importance in the frequency range from about 1 GHz to 1 000 GHz where experimental techniques change gradually from standard radio-circuits to optical systems. They also serve to remind us of the existence of types of wave appreciably more complicated than simple plane waves. Guided waves have many properties which are quite different from those of plane waves, thus, for example, the guided waves used in travelling-wave tubes have strong components of the fields in the direction of propagation and both the phase and group velocities are much less than c.

Diffraction

If a plane wave propagating along the z-axis as $\exp\{\mathrm{j}(\omega t-(\omega/c)z)\}$ traverses a diffraction grating, or transparent object such as a photographic plate, in the xy plane, the obstacle will modulate the amplitude of the light wave and each component of the field in the xy plane will be a function of x and y. This function can be expressed as a Fourier integral or series, and detail in the object on a scale x_0 in the x-direction will lead to Fourier components which vary as $\exp\{-2\pi\mathrm{j}(x/x_0)\}$ and $\exp\{+2\pi\mathrm{j}(x/x_0)\}$. Because the fields just beyond the object must be continuous these components will launch waves which vary with t, x and z as $\exp\{\mathrm{j}(\omega t\mp 2\pi(x/x_0)-\beta' z)\}$, and because each field component must satisfy the wave equation, we have $(2\pi/x_0)^2+\beta'^2 = \omega^2/c^2$. But a wave whose propagation constant is $\boldsymbol{\beta} = (\pm 2\pi/x_0, 0, \beta')$ propagates at an angle θ to the z-axis where $\tan\theta = \pm 2\pi/\beta' x_0$ or $\sin\theta = \pm 2\pi c/\omega x_0 = \pm\lambda/x_0$. Thus detail on a scale x_0 generates diffracted beams which propagate away from the object at angles $\pm\sin^{-1}(\lambda/x_0)$. This is the fundamental process involved in the diffraction of light. If the light leaving the object is collected by a lens of angular aperture θ, information about detail on a scale x_0 will only be collected if $\sin\theta > \lambda/x_0$ or $x_0 > \lambda/\sin\theta$. This is the basic formula determining the resolving power of an optical instrument.

PROBLEMS

10.1. Verify eqns (10.2) and (10.3) of the text.

10.2. In a.c. circuit theory a real physical voltage is often represented as a complex quantity $V\exp(\mathrm{j}\omega t)$. Explain the meaning of this notation. If the current in an a.c. circuit is $I\exp(\mathrm{j}\omega t)$ and the voltage across it is $V\exp(\mathrm{j}\omega t)$ show from first principles that the average power dissipated in the circuit is the real part of $\frac{1}{2}IV^*$.

10.3. Verify eqn (10.5) of the text.

10.4. A vacuum-spaced coaxial line has inner and outer conductors of radii a and b. Calculate L and C and show that $LC = \mu_0\epsilon_0$. Calculate the characteristic impedance Z_0 and find its numerical value if $b = 3a$.

10.5. Verify eqn (10.11) of the text.

10.6. Discuss the input impedance of a line of characteristic impedance Z_0 terminated in a load Z for the following cases. (a) $Z = Z_0$. (b) $\beta l = n\pi$ where n is integral. (c) $Z = 0$ and $\beta l = \pi/2$. (d) $Z = \infty$ and $\beta l = \pi/2$. (e) $Z = jZ_0$ and $\beta l = \pi/8$. (f) $Z = 0$ and βl is small but finite. (g) $Z = \infty$ and βl is small but finite.

10.7. Alternate sections of line, of equal lengths l but different impedances Z_1 and Z_2, are connected together. There are n sections of each type and the last section of impedance Z_2 is connected to a load resistance $R = Z_0$. Calculate the input impedance (a) when $\beta l = \pi$ and (b) when $\beta l = \pi/2$. How could this structure be used as a filter? If the input is connected to a line of impedance Z_0 what is the reflection coefficient at the junction when $\beta l = \pi/2$, $n = 10$, $Z_1 = \frac{1}{2}Z_0$ and $Z_2 = \frac{2}{3}Z_0$?

10.8. The conductors in a coaxial line are held in position by dielectric spacers. Show that the reflection due to each spacer is a minimum when the spacer is half a wavelength long and a maximum when it is a quarter-wavelength long.

10.9. Verify that $[\nabla \wedge (\nabla \wedge \boldsymbol{E})]_x = (\partial/\partial x)(\nabla . \boldsymbol{E}) - \nabla^2 E_x$.

10.10. Verify eqns (10.14a) and (10.14b) of the text.

10.11. Stefan's law gives the power radiated by a black body at a temperature T as $6(T/100)^4$ watts m^{-2}. What is the r.m.s. electric field just outside the surface of a body at room temperature?

10.12. Show that when light falls at normal incidence on a glass surface approximately 4 per cent of the intensity is reflected. A prism-binocular optical system contains at least six air–glass interfaces. How much light is lost by reflection at these surfaces?

10.13. Unpolarized white light falls on a glass surface at Brewster's angle. Show that the reflected light is polarized and calculate the fraction of the incident intensity in the reflected ray. Assume that the refractive index is 1·5.

10.14. The conductivity of copper is $6{\cdot}10^7$ mho m^{-1}. Compare the resistance of a round wire of diameter 1 mm at 50 MHz with its d.c. resistance.

10.15. Bus bars in a power station working at 50 Hz are usually constructed from several copper bars each $\frac{1}{4}''$ thick connected in parallel. Why is this?

10.16. What is the reflectivity of silver at the wavelength of red light, 6 000 Å? The conductivity is 6×10^7 mho m^{-1}.

10.17. The ionosphere reflects radio-waves of frequency up to 20 MHz. Estimate the electron density in the ionosphere.

10.18. Light propagates in a transparent rod or light pipe and is confined to the light pipe if the angle of incidence of the rays at the surface is large enough to lead to total internal reflection. Find the phase velocity of the waves as a function of the angle of incidence and show that, at the critical angle, it is equal to c.

10.19. A rectangular waveguide has internal dimensions $1'' \times \frac{1}{2}''$. What is the lowest frequency that will propagate in this guide?

10.20. The power flowing in the guide of problem 10.19 is 1 kwatt at 10^4 MHz. What is the value of the peak electric field?

10.21. A grating with equal clear and opaque lines of width a is illuminated at normal incidence by light of wavelength $\lambda > 2a$. Show that there are no diffracted rays in the transmitted light and that the intensity of the transmitted light is reduced by $\frac{1}{4}$.

10.22. A dielectric mirror consists of alternate layers of two media of refractive indices $n_1 = 2{\cdot}4$ and $n_2 = 1{\cdot}4$. Each layer is a quarter of a wavelength thick and there are $N = 20$ layers of each type. Calculate the reflectivity for light at normal incidence (see problem 10.7).

10.23. A coaxial line with an outer conductor of radius 1 cm has an inner conductor whose radius changes suddenly from 0·5 cm to 0·1 cm. Calculate the reflection coefficient due to this discontinuity.

10.24. If $\boldsymbol{E}$ is expressed in terms of its Cartesian components the result of problem (10.9) leads to $\nabla \wedge (\nabla \wedge \boldsymbol{E}) = \nabla(\nabla . \boldsymbol{E}) - \nabla^2 \boldsymbol{E}$. Why is this not a valid result if $\boldsymbol{E}$ is expressed in terms of, say, spherical polar components?

10.25. The electric field associated with a linearly polarized plane wave in vacuum is $E_y = E_0 \exp \mathrm{j}(\omega t - \boldsymbol{\beta} . \boldsymbol{r})$ where the components of $\boldsymbol{\beta}$ are $\omega/c(\sin\theta, 0, \cos\theta)$. What is the direction of propagation of the wave and what is the distance between the intercepts of the wave fronts on the plane $x = 0$?

10.26. Two plane waves similar to that in question (10.25) but with values of $\theta = \theta_1$ and $-\theta_1$ are superimposed. Show that E_y vanishes in the planes

$$x = \pm \frac{\pi c}{2\omega \sin\theta}.$$

Find the components of the associated magnetic intensity $\boldsymbol{H}$ and compare your results with eqns (10.32), (10.33), and (10.34).

10.27. Show that a body for which ϵ is negative at some frequency ω is totally reflecting at that frequency.

10.28. Combine the results of problems (10.27) and (8.15) to find the range of frequencies over which a crystal with an optical refractive index $n = 1{\cdot}5$, a static dielectric constant $\epsilon(0) = 6$ and a single sharp ionic resonance at $3{\cdot}10^{13}$ Hz is totally reflecting.

11. Radiation

To calculate the power radiated by accelerated charges or circulating currents in systems such as atoms or radio aerials we need an expression for the distant fields due to the systems so that we can calculate the total outward flux of energy from Poynting's vector $\boldsymbol{E} \wedge \boldsymbol{H}$. Although it is possible to calculate the fields directly it is much easier to proceed indirectly and begin by introducing the scalar and vector potentials. If we write $\boldsymbol{B} = \boldsymbol{\nabla} \wedge \boldsymbol{A}$ and $\boldsymbol{E} = -\dot{\boldsymbol{A}} - \boldsymbol{\nabla}\phi$ the two equations $\boldsymbol{\nabla}.\boldsymbol{B} = 0$ and $\boldsymbol{\nabla} \wedge \boldsymbol{E} + \dot{\boldsymbol{B}} = 0$ are identically satisfied. The equation $\boldsymbol{\nabla} \wedge \boldsymbol{B} - \mu_0\epsilon_0\dot{\boldsymbol{E}} = \mu_0\boldsymbol{J}$ then becomes

$$\boldsymbol{\nabla} \wedge \boldsymbol{\nabla} \wedge \boldsymbol{A} + \mu_0\epsilon_0\ddot{\boldsymbol{A}} = \mu_0\boldsymbol{J} - \mu_0\epsilon_0\,\boldsymbol{\nabla}\dot{\phi}$$

and $\boldsymbol{\nabla}.\boldsymbol{E} = \rho/\epsilon_0$ becomes $\nabla^2\phi = -\rho/\epsilon_0 - \boldsymbol{\nabla}.\dot{\boldsymbol{A}}$. We can rearrange these equations as

$$\nabla^2\boldsymbol{A} - \mu_0\epsilon_0\ddot{\boldsymbol{A}} = -\mu_0\boldsymbol{J} + \boldsymbol{\nabla}(\mu_0\epsilon_0\dot{\phi} + \boldsymbol{\nabla}.\boldsymbol{A}) \tag{11.1a}$$

and

$$\nabla^2\phi - \mu_0\epsilon_0\ddot{\phi} = -\frac{\rho}{\epsilon_0} - \frac{\partial}{\partial t}(\mu_0\epsilon_0\dot{\phi} + \boldsymbol{\nabla}.\boldsymbol{A}) \tag{11.1b}$$

as long as we keep to Cartesian coordinates. If we add to $\boldsymbol{A}$ the gradient of a scalar $\boldsymbol{\nabla}\psi$ it leaves $\boldsymbol{B}$ unchanged but $\boldsymbol{E}$ changes by $-\boldsymbol{\nabla}\dot{\psi}$. Thus if

$$\boldsymbol{A} \to \boldsymbol{A}_0 = \boldsymbol{A} + \boldsymbol{\nabla}\psi \quad \text{and} \quad \phi \to \phi_0 = \phi - \dot{\psi}$$

both $\boldsymbol{E}$ and $\boldsymbol{B}$ remain unchanged and $\boldsymbol{A}_0$ and ϕ_0 are equally acceptable potentials. The expression in brackets on the right of (11.1a) and (11.1b) is changed by $\nabla^2\psi - \mu_0\epsilon_0\ddot{\psi}$ and, by a proper choice of ψ, we can arrange that this just cancels the expression completely. This is equivalent to saying that we can always choose potentials $\boldsymbol{A}$ and ϕ which give the correct fields and also satisfy $\mu_0\epsilon_0\dot{\phi} + \boldsymbol{\nabla}.\boldsymbol{A} = 0$. These Lorentz potentials satisfy the simpler equations

$$\nabla^2\boldsymbol{A} - \mu_0\epsilon_0\ddot{\boldsymbol{A}} = -\mu_0\boldsymbol{J} \tag{11.2a}$$

$$\nabla^2\phi - \mu_0\epsilon_0\ddot{\phi} = -\rho/\epsilon_0 \tag{11.2b}$$

and equations of this form can be shown to have solutions

$$\boldsymbol{A}(\boldsymbol{r}_1, t_1) = \int_V \frac{\mu_0\boldsymbol{J}(\boldsymbol{r}_2, t_2)\,\mathrm{d}V(\boldsymbol{r}_2)}{4\pi r_{12}} \tag{11.3a}$$

and

$$\phi(\boldsymbol{r}_1, t_1) = \int_V \frac{\rho(\boldsymbol{r}_2, t_2)\,\mathrm{d}V(\boldsymbol{r}_2)}{4\pi\epsilon_0 r_{12}}, \tag{11.3b}$$

where

$$r_{12} = |\boldsymbol{r}_1 - \boldsymbol{r}_2| \tag{11.4}$$

and

$$t_2 = t_1 - (\mu_0\epsilon_0)^{\frac{1}{2}} r_{12}. \tag{11.5}$$

These solutions are known as the retarded potentials. They express the potentials, at a point $\boldsymbol{r}_1$ at a time t_1, in terms of the current and charge in a volume element dV at $\boldsymbol{r}_2$ at an earlier time $t_2 = t_1 - (r_{12}/c) = t_1 - (\mu_0\epsilon_0)^{\frac{1}{2}} r_{12}$. They express, in the most general way, the fact that an electromagnetic disturbance takes a time r_{12}/c to propagate from $\boldsymbol{r}_2$ to $\boldsymbol{r}_1$.

Consider now a localized set of charges and currents near the origin: then to calculate the power radiated we need the fields on a large sphere of radius R surrounding the origin and the power radiated is given by

$$W = \int_0^{2\pi} \int_0^{\pi} \frac{1}{\mu_0} (\boldsymbol{E} \wedge \boldsymbol{B})_R R^2 \sin\theta \, \mathrm{d}\theta \, \mathrm{d}\varphi. \tag{11.6}$$

Obviously only components of $\boldsymbol{E}$ and $\boldsymbol{B}$ which are tangential to the sphere contribute to $(\boldsymbol{E} \wedge \boldsymbol{B})$. Further only components which fall off no faster than $1/R$ give a finite contribution. There are no such components in $-\nabla\phi$ and so in dealing with radiation we can set $\boldsymbol{E} = -\dot{\boldsymbol{A}}$. Furthermore, at a distance, the fields behave like plane waves and the tangential component of $(1/\mu_0)\boldsymbol{B}$ is $(\epsilon_0/\mu_0)^{\frac{1}{2}}$ times that of $\boldsymbol{E}$. Thus, to this approximation, we need only calculate $\dot{\boldsymbol{A}}$. In Fig. 11.1 a particle of charge q moves with a velocity $\boldsymbol{v}$ past the origin of a system of polar axes and, in (11.3a) the volume integral of $\boldsymbol{J}$ is simply $\int \boldsymbol{J}\,\mathrm{d}V = q\boldsymbol{v}$ so that $\boldsymbol{A} = \mu_0 q\boldsymbol{v}/4\pi R$ and the only tangential component is $A_\theta = -\mu_0 qv \sin\theta/4\pi R$. The electric-field component tangential to the sphere

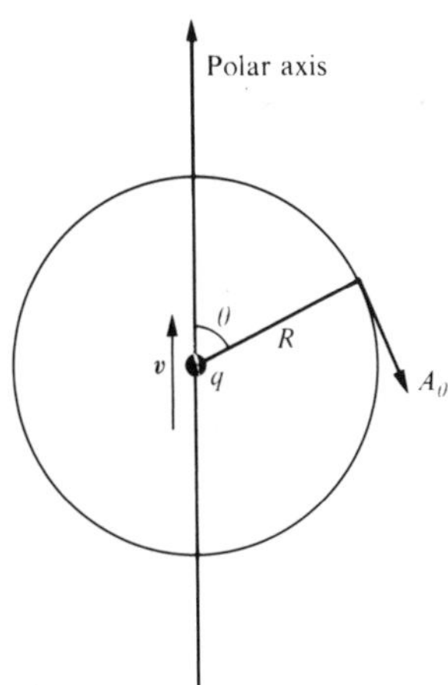

FIG. 11.1. A charge q moving along the polar axis with a velocity $\boldsymbol{v}$.

is

$$E_\theta = -\dot{A}_\theta = \frac{\mu_0 q\dot{v}\sin\theta}{4\pi R}$$

and

$$(\boldsymbol{E}\wedge\boldsymbol{H})_R = \frac{1}{\mu_0}(\boldsymbol{E}\wedge\boldsymbol{B})_R = E_\theta H_\varphi = \left(\frac{\epsilon_0}{\mu_0}\right)^{\frac{1}{2}}\frac{\mu_0^2 q^2\dot{v}^2\sin^2\theta}{(4\pi R)^2}. \tag{11.7}$$

We see that the power radiated is proportional to the square of the acceleration $\dot{v}$ and a maximum at $\theta = \pi/2$ i.e. at right angles to the direction of motion of the charge. Eqn (11.7) can be rearranged as

$$(\boldsymbol{E}\wedge\boldsymbol{H})_R = \left(\frac{\mu_0}{\epsilon_0}\right)^{\frac{1}{2}}\frac{q^2\dot{v}^2\sin^2\theta}{(4\pi cR)^2} \tag{11.8}$$

and the total power radiated obtained from (11.6) and (11.8) is

$$W = \frac{2}{3}\left(\frac{\mu_0}{\epsilon_0}\right)^{\frac{1}{2}}\frac{q^2\dot{v}^2}{4\pi c^2}. \tag{11.9}$$

This formula, applied to electrons, appears continually in atomic physics.

If a charge q oscillates with a frequency $\omega/2\pi$ and maximum displacement l about a fixed centre it represents an oscillatory dipole of moment $p = ql$ and we have $q\dot{v} = -\omega^2 ql$ so that

$$W = \frac{1}{3}\left(\frac{\mu_0}{\epsilon_0}\right)^{\frac{1}{2}}\frac{\omega^4 p^2}{4\pi c^2} \tag{11.10}$$

where the extra factor $\frac{1}{2}$ comes from taking the time average. This formula can also be applied to atomic electrons but, if we have a short dipole at radio-frequencies, $\dot{p} = \dot{q}l = Il$ where I is the current feed to the dipole and so

$$W = \frac{1}{3}\left(\frac{\mu_0}{\epsilon_0}\right)^{\frac{1}{2}}\frac{I^2\omega^2 l^2}{4\pi c^2} = \frac{1}{12\pi}\left(\frac{\mu_0}{\epsilon_0}\right)^{\frac{1}{2}} I^2\left(\frac{2\pi l}{\lambda}\right)^2. \tag{11.11}$$

If we write the power radiated as $W = \frac{1}{2}RI^2$ with

$$R = \frac{1}{6\pi}\left(\frac{\mu_0}{\epsilon_0}\right)^{\frac{1}{2}}\left(\frac{2\pi l}{\lambda}\right)^2 \sim 20\left(\frac{2\pi l}{\lambda}\right)^2 \text{ ohms} \tag{11.12}$$

the quantity R is known as the radiation resistance and the voltage applied to drive the dipole contains a component RI in phase with I in addition to the out-of-phase or quadrature component needed to charge the dipole. This formula is only reasonably accurate if l is less than about $\lambda/4$.

PROBLEMS

11.1. Show that the vector potentials $\boldsymbol{A}_1 = (-y, x, 0)$ and $\boldsymbol{A}_2 = (0, 2x, 0)$ lead to the same field $\boldsymbol{B}$. Calculate $\nabla . \boldsymbol{A}_1$ and $\nabla . \boldsymbol{A}_2$.

11.2. In eqn (11.3b) divide the region of integration into (a) a small sphere of radius $\alpha \to 0$ about the point $\boldsymbol{r}_1$ and (b) the rest of space. Show that the contribution to $\nabla^2\phi$ from the interior of the sphere is $-\rho/\epsilon_0$ and the contribution from the outside region is $+\mu_0\epsilon_0\ddot{\phi}$. Use this to verify that (11.3b) satisfies (11.2b). (See Abraham and Becker (1932).)

11.3. An electron oscillates along a line of length 1 angström at a frequency of 10^{15} Hz. Calculate the rate at which it loses energy. Express your answer in electronvolts per second and relate this to the kinetic energy of the electron also expressed in electronvolts.

11.4. A short dipole aerial of length 20 cm is fed at the centre at 100 MHz and the capacitance between the two halves of the dipole is 1·5 pF. Calculate the radiation resistance and the current needed to radiate 1 watt. What voltage is required to drive the aerial?

11.5. Plane waves polarized with $\boldsymbol{E}$ parallel to the x-axis are scattered by atoms in which the electronic resonances lie well above the frequency of the radiation. Show (i) that the light is predominantly scattered at right angles to the x-axis and (ii) that the scattering is proportional to the fourth power of the frequency of the radiation. How is this result modified for light scattered by free electrons in a plasma?

11.6. A cloud contains spherical rain-drops of radius 10^{-7} m and refractive index 1·33. The number of rain drops per cubic metre is 10^{10}. A parallel beam of laser radiation of wavelength 10^{-5} m and intensity W watts m^{-2} traverses the cloud. Calculate the induced dipole moment of a single droplet and use this result to find the intensity of the radiation scattered at right angles to the initial beam.

11.7. In systems in which $\rho = 0$ the scalar potential ϕ can be omitted so that $\boldsymbol{E} = -\dot{\boldsymbol{A}}$, and, if the currents vary slowly the effects of retardation can also be ignored so that t_2 can be set equal to t_1 in eqn (11.3a). Use these results to derive eqn (3.10) (Neumann's formula).

11.8. The function $G(\theta, \varphi)$, which gives the fraction of the total power radiated by an aerial going into a solid angle $\mathrm{d}\Omega$ at the polar angles θ, φ, as $G(\theta, \varphi)\,\mathrm{d}\Omega$ is known as the aerial gain. Find the form of $G(\theta, \varphi)$ for a short dipole.

11.9. The mean-square, open-circuit voltage developed at the terminals of an aerial of radiation resistance R_r and gain $G(\theta, \varphi)$ when a plane wave of the correct polarization and intensity N falls on it from the direction θ, φ is $4\lambda^2 R_r N G(\theta, \varphi)$ where λ is the wavelength. Use this result to derive the r.m.s. voltage at the terminals of a short dipole of length $l \ll \lambda$ with its axis parallel to the electric vector (of r.m.s. amplitude E) of a plane wave.

Bibliography

THE books in this list represent only a small fraction of the total number of available texts. They have been chosen to supplement the material in the text and to introduce the reader to the standard literature on the subject.

1. A. F. KIP *Fundamentals of electricity and magnetism* McGraw–Hill (1969) is at about the same level as the present text but the emphasis is different.
2. B. BLEANEY and B. BLEANEY *Electricity and magnetism* Clarendon Press (1965) is a longer book giving a much fuller coverage, especially of applications and the experimental aspects of the subject.
3. J. D. JACKSON *Classical electrodynamics* Wiley (1965) is a much more advanced theoretical text. The treatment of radiation and those aspects of electromagnetism which find applications in nuclear and atomic physics is notably thorough and clear.
4. W. PANOFSKY and M. PHILIPS *Classical electricity and magnetism* Addison–Wesley (2nd ed. 1962) is similar to 3.
5. S. RAMO, J. R. WHINNERY, and T. VAN DUZER *Fields and waves in electronic communications* Wiley (1965) is a relatively advanced but clearly written text on the applications of electromagnetic theory.
6. S. A. SCHELKUNOFF *Electromagnetic waves* Van Nostrand (1943) is a more formal text and a source book for this aspect of the subject.
7. M. ABRAHAM and R. BECKER *Classical electricity and magnetism* Blackie (1932) is a less advanced text than 3, 4, 5, and 6 but it is notable for the full discussion it gives of mathematical methods.
8. L. D. LANDAU and E. M. LIFSHITZ *Electrodynamics of continuous media* Pergamon (1960) is a difficult but quite outstanding text.
9. F. N. H. ROBINSON *Macroscopic electromagnetism* Pergamon Press, Oxford (1973) Deals, at an advanced level, with the topics of chapters 6, 7, 8, and 9.
10. C. KITTEL *Introduction to solid-state physics* Wiley (1971) is a useful introduction to the relation between electromagnetic properties and the structure of matter.
11. H. FROHLICH *Theory of Dielectrics* Clarendon Press (1958) is more specialized but not difficult.
12. J. H. VAN VLECK *Electric and magnetic susceptibilities* Clarendon Press (1932). Because of its age parts of this book are rather dated but nevertheless it remains much the best advanced text in this field. The treatment of the fundamental atomic phenomena involved in dielectric and magnetic properties is uniquely thorough.
13. H. A. LORENTZ *The theory of electronics* Dover (1952). Originally published in 1908 this book is one of the classics of electromagnetism.
14. L. ROSENFELD *Theory of electrons* Dover (1966) treats many of the same topics as Lorentz and is rather easier reading.
15. E. T. WHITTAKER *History of the theories of the aether and electricity* Nelson 2 volumes (1953) is a mine of information on many neglected aspects of electromagnetism and, because its author was a distinguished mathematical physicist, is almost unique, amongst historical works, in dealing adequately with the physical content of the subject.
16. J. H. JEANS *Electricity and magnetism* Cambridge University Press (1951).

17. J. A. STRATTON *Electromagnetic theory* McGraw–Hill (1941).
18. W. R. SMYTHE *Static and dynamic electricity* McGraw–Hill (1968) are three texts which though difficult to read contain a wealth of invaluable information and solved problems.

Finally we remind the reader that electromagnetic properties are tabulated. New volumes in the series (Landolt–Bornstein) published by Springer continue to appear and from time to time accounts of developments in the measurement of the fundamental electromagnetic constants appear in the 'Reviews of Modern Physics' published by the American Physical Society.

Index

Some useful relations in vector analysis

If ϕ and Ψ are scalars and $\boldsymbol{F}$ a vector:

$$\operatorname{div}(\phi \boldsymbol{F}) = \boldsymbol{F} \cdot \mathbf{grad}\, \phi + \phi \operatorname{div} \boldsymbol{F}$$
$$\mathbf{curl}(\phi \boldsymbol{F}) = \phi\, \mathbf{curl}\, \boldsymbol{F} - \boldsymbol{F} \wedge \mathbf{grad}\, \phi$$

$$\iiint (\Psi \nabla^2 \phi - \phi \nabla^2 \Psi)\, \mathrm{d}V = \iint (\Psi\, \mathbf{grad}\, \phi - \phi\, \mathbf{grad}\, \Psi) . \mathrm{d}\boldsymbol{S}$$

The Laplacian Operator ∇^2

Cartesian coordinates:

$$\nabla^2 \Psi = \frac{\partial^2 \Psi}{\partial x^2} + \frac{\partial^2 \Psi}{\partial y^2} + \frac{\partial^2 \Psi}{\partial z^2}$$

Cylindrical coordinates:

$$\nabla^2 \Psi = \frac{1}{r} \frac{\partial}{\partial r}\left(r \frac{\partial \Psi}{\partial r} \right) + \frac{1}{r^2} \frac{\partial^2 \Psi}{\partial \varphi^2} + \frac{\partial^2 \Psi}{\partial z^2}$$

Spherical polar coordinates:

$$\nabla^2 \Psi = \frac{1}{r^2} \frac{\partial}{\partial r}\left(r^2 \frac{\partial \Psi}{\partial r} \right) + \frac{1}{r^2 \sin \theta} \frac{\partial}{\partial \theta}\left(\sin \theta \frac{\partial \Psi}{\partial \theta} \right) + \frac{1}{r^2 \sin^2 \theta} \frac{\partial^2 \Psi}{\partial \varphi^2}$$

$\mathbf{curl}\ \mathbf{curl}\ \boldsymbol{F} = \mathbf{grad} \operatorname{div} \boldsymbol{F} - \nabla^2 \boldsymbol{F}$, where $\nabla^2 \boldsymbol{F}$ is to be interpreted as

$$\nabla^2 \boldsymbol{F} = \hat{\boldsymbol{x}} \nabla^2 F_x + \hat{\boldsymbol{y}} \nabla^2 F_y + \hat{\boldsymbol{z}} \nabla^2 F_z$$

and $\hat{\boldsymbol{x}}$, $\hat{\boldsymbol{y}}$, $\hat{\boldsymbol{z}}$ are unit vectors in the x-, y-, and z-directions

Physical constants and conversion factors

Avogadro constant	L or N_A	$6{\cdot}022 \times 10^{23}$ mol^{-1}
Bohr magneton	μ_B	$9{\cdot}274 \times 10^{-24}$ J T^{-1}
Bohr radius	a_0	$5{\cdot}292 \times 10^{-11}$ m
Boltzmann constant	k	$1{\cdot}381 \times 10^{-23}$ J K^{-1}
charge of an electron	e	$-1{\cdot}602 \times 10^{-19}$ C
Compton wavelength of electron	$\lambda_C = h/m_e c$	$= 2{\cdot}426 \times 10^{-12}$ m
Faraday constant	F	$9{\cdot}649 \times 10^{4}$ C mol^{-1}
fine structure constant	$\alpha = \mu_0 e^2 c/2h$	$= 7{\cdot}297 \times 10^{-3}$ ($\alpha^{-1} = 137{\cdot}0$)
gas constant	R	$8{\cdot}314$ J K^{-1} mol^{-1}
gravitational constant	G	$6{\cdot}673 \times 10^{-11}$ N m^2 kg^{-2}
nuclear magneton	μ_N	$5{\cdot}051 \times 10^{-27}$ J T^{-1}
permeability of a vacuum	μ_0	$4\pi \times 10^{-7}$ H m^{-1} exactly
permittivity of a vacuum	ϵ_0	$8{\cdot}854 \times 10^{-12}$ F m^{-1} ($1/4\pi\epsilon_0 = 8{\cdot}988 \times 10^{9}$ m F^{-1})
Planck constant	h	$6{\cdot}626 \times 10^{-34}$ J s
(Planck constant)/2π	$\hbar$	$1{\cdot}055 \times 10^{-34}$ J s $= 6{\cdot}582 \times 10^{-16}$ eV s
rest mass of electron	m_e	$9{\cdot}110 \times 10^{-31}$ kg $= 0{\cdot}511$ MeV/c^2
rest mass of proton	m_p	$1{\cdot}673 \times 10^{-27}$ kg $= 938{\cdot}3$ MeV/c^2
Rydberg constant	$R_\infty = \mu_0^2 m_e e^4 c^3/8h^3$	$= 1{\cdot}097 \times 10^{7}$ m^{-1}
speed of light in a vacuum	c	$2{\cdot}998 \times 10^{8}$ m s^{-1}
Stefan–Boltzmann constant	$\sigma = 2\pi^5 k^4/15h^3c^2$	$= 5{\cdot}670 \times 10^{-8}$ W m^{-2} K^{-4}
unified atomic mass unit (^{12}C)	u	$1{\cdot}661 \times 10^{-27}$ kg $= 931{\cdot}5$ MeV/c^2
wavelength of a 1 eV photon		$1{\cdot}243 \times 10^{-6}$ m

1 Å $= 10^{-10}$ m; 1 dyne $= 10^{-5}$ N; 1 gauss (G) $= 10^{-4}$ tesla (T);
0°C $= 273{\cdot}15$ K; 1 curie (Ci) $= 3{\cdot}7 \times 10^{10}$ s^{-1};
1 J $= 10^{7}$ erg $= 6{\cdot}241 \times 10^{18}$ eV; 1 eV $= 1{\cdot}602 \times 10^{-19}$ J; 1 cal$_{th}$ $= 4{\cdot}184$ J;
ln 10 $= 2{\cdot}303$; ln $x = 2{\cdot}303 \log x$; e $= 2{\cdot}718$; log e $= 0{\cdot}4343$; $\pi = 3{\cdot}142$